DEVIL'S TANGO

Other books by Cecile Pineda:

Face
Frieze
The Love Queen of the Amazon
Fishlight: A Dream of Childhood
Bardo99: A Mononovel
Redoubt: A Mononovel

DEVIL'S TANGO

HOW I LEARNED THE FUKUSHIMA STEP BY STEP

Cecile Pineda

with a Foreword by Jacqueline Cabasso
Executive Director, Western States Legal Foundation

San Antonio, Texas

Devil's Tango: How I Learned the Fukushima Step by Step

Second (Revised and Expanded) Edition
March 2013

Print Edition ISBN: 978-1-60940-315-7
ePub ISBN: 978-1-60940-316-4
Kindle ISBN: 978-1-60940-317-1
PDF ISBN: 978-1-60940-318-8

Wings Press
627 E. Guenther
San Antonio, Texas 78210
Phone/fax: (210) 271-7805

On-line catalogue and ordering:
www.wingspress.com
All Wings Press titles are distributed to the trade by
Independent Publishers Group
www.ipgbook.com

Library of Congress Cataloging-in-Publication Data (first edition):

Pineda, Cecile.
Devil's tango : how I learned the Fukushima step by step / Cecile Pineda. -- 1st ed.
p. cm.
Includes bibliographical references.
ISBN 978-0-916727-99-4 (pbk. : alk. paper) -- ISBN 978-1-60940-234-1 (ePub eBook) -- ISBN 978-1-60940-235-8 (Kindle eBook) -- ISBN 978-1-60940-236-5 (library PDF eBook)
1. Nuclear power plants--Risk assessment. 2. Fukushima Nuclear Disaster, Japan, 2011. 3. Environmental disasters--Forecasting. 4. Overpopulation. I. Title.
TK9152.16.P56 2012
363.17'992--dc23

2012004108

Contents

Foreword

Sixty-seven years into the nuclear age, we are all Fukushima now: Hiroshima, Nagasaki, Bikini, Three-Mile Island, Chernobyl, uranium mining, depleted uranium, cancers, leukemias, extreme birth defects, and the list goes on.... How has the human race backed itself into this radioactive "End Game," and what can be done to keep the earth from becoming a designated exclusion zone? Told in the first person, *Devil's Tango: How I Learned the Fukushima Step by Step* is an extraordinary and fearless slice through the meaning of life—and death—from the very personal to the very global. The unfolding Fukushima nuclear disaster, exhaustively researched, serves as a day-to-day benchmark, calling into question the very future of our mother planet. Profound, quirky, laugh-out-loud funny at times, and utterly terrifying, the author challenges us to join her in waking ourselves out of denial, in learning to see the connections that link the tiniest fish scale and the smallest grain of sand with the corrupt juggernaut of the corporate, international terror state, and in acting on our consciences. It's up to us to save ourselves. Change, the author concludes, won't come from governments or the halls of power. It will come from the 99% and from activists and local people around the world in a life-and-death, decades-long struggle for the sanctity of life. *Devil's Tango* tells an amazing story. Read it if you dare. Read it if you care.

Jacqueline Cabasso
Executive Director
Western States Legal Foundation
February 5, 2012

Preface to the First Edition

Devil's Tango did not teach me to dance. It started with a total car wreck from which I escaped unscathed except for a bloody nose and a deviated septum. I didn't want the wreckage to remind me: I couldn't get it towed to the junkyard fast enough. It ended with a broken ankle on New Year's Eve. You may ask was I boozing, cutting too many quicksteps, but I hate to disappoint. I was hiking. I "let go" after nine months of feverish work learning this tango's numerous steps. I was *not paying attention.*

It's not easy for you, or me, or anyone to *pay attention* to the consequences of the nuclear energy cycle. Why? Because you can't see radiation. And you can't tell if your cancer happened because ten years earlier you were in a place it would have been better for you not to be. You can't tell if your stillbirth happened because your tiny fetus couldn't live in an irradiated world. And you can't tell if the child born to you 10, 25, 50 years ago, and who is a "different" child, affected by autism, or ADD, or any number of unidentified syndrome complexes, was born different because the air was contaminated from "testing," or because the milk that year was tainted from fallout.

You can't see fallout, you can't tell when you're eating strontium by the spoonful. It's invisible, you can't see it, feel it, touch it, hear it: you can taste it only sometimes—when the fallout is particularly dense—as a metallic taste in your mouth, which any number of people reported this past year in places as far apart as Seattle and Arizona. In a world that enshrines surfaces, the industry thinks invisibility is a sure bet you won't ever find out.

To say that corporate enterprise has abrogated your right to ask questions, to raise objections, or to expose malfeasance tells only half the story. Corporations are not people. We are people, and until we learn to protest *en masse,* until we make it impossible for corporations to continue stripping the planet, they will hijack our earth, and make all living things expendable.

That means you. That means us.

Writing this book taught me that I was every bit as asleep as anyone else. It woke me up. I hope that in reading it, you will want to join me, in the streets, in your town hall, in your council chambers, at your local reactor's gate, and wherever like-minded people bent on defending the left-over tatters of our planet meet. Abbie Hoffman once wrote: "Democracy is not something you believe in or hang your hat on, but something you *do*. You participate. If you stop doing it, democracy stumbles and falls. If you participate, the future is yours."

Cecile Pineda
Berkeley, California
January 30, 2012

Preface to the Second Edition*

How Do We Live?

The question, "How do we live knowing what we know?" is the question, more than any other, that people all across the United States have brought up during the dozens of *Devil's Tango* presentations I have made in the year since the book's publication on March 11, 2012. During that time I have addressed thousands of people, many of whom live in the shadows of old, embrittling reactors, two of them the same defective GE Mark I BWRs that failed at Fukushima when they succumbed to the stresses of R-9 and R-6.4 earthquakes on March 11 and 12, 2011.

We have always lived in the knowledge of mutability. The only thing we can count on to remain the same is change itself. But now we have begun to live in a quantum shift. We can see the looming destruction of our home, of our planet and of all living things from the moment that the divine sparking of life on earth began. And yet, we have a choice, and choice is what ultimately lends dignity to human life. No matter how little may seem possible in the real world—the world out there—in the human spirit, everything is possible.

It is appropriate to look at the events which began on March 11, 2011—events still unfolding even now—as a planet-threatening catastrophe. We are all well advised to understand their appalling dimensions, but at the same time, if we allow ourselves to really see, we may discover in them a critical invitation to become even more who we were meant to become by extending ourselves even more to

---

* The second edition, revised and expanded, provides more accurate information about depleted uranium. U-235 enrichment is required both to make nuclear weapons (bombs and "bunker busters") and to fuel nuclear reactors. Depleted uranium, DU, is the by-product of that enrichment process. Based on signature evidence, only very minor quantities—if any—of nuclear waste are used in the manufacture of depleted uranium.

our loved ones and to every human being, to every form of life on earth, expressing our generosity, our ability to honor, and respect, and share with one another; and by taking a very public stand in the face of an economic system that, in its blind rapacity, has taken upon itself such Godzilla proportions that it threatens to gut the very planet on which we live.

As you allow yourself to hear the many voices speaking to you in this book, may you feel the invitation to choose to engage. Against the paralysis of fear and despair, I know of no better defense.

Cecile Pineda
Berkeley, California
February 1, 2013

DEVIL'S TANGO

HOW I LEARNED THE FUKUSHIMA STEP BY STEP

1.
Habitable Zones

We picked out planets that are just the right size—between the size of Earth or twice that—and all are within the 'habitable zones' of their stars, at distances where there's the best chance for liquid water—and possibly life—to exist.

— Dan Wertheimer
Space sciences astrophysicist

There is no place more wonderful than this. There is no place more marvelous than here.

— Milarepa

Starry night. All along the horizon, telescopes rotate, staring at the night sky: the Atacama Desert, where the skies are transparent like no other place on earth, free of the pollution of city lights, and of temperate zone moisture.

The human race is looking for planets. Hungry for planets in our own image, in the image of Gaia, of Earth. Planets near enough yet far enough from their distant suns not to burn up, not to freeze. Planets which show signs of water in their atmospheres. Planets that revolve around the maybe 50 billion stars in the local galaxy, in the neighborhood we call the Milky Way, and in the narrowest possible tranche of it, 1,235 planets have been sighted that correspond to such spacial parameters, and of those 1,235, 86 stand out, 86 which answer within reasonable limits to those conditions: sufficiently distant from their suns (but not too distant) to entertain the possibility of water.

Imagine 86 watery planets, each with its own orders of life: its own set of one-celled organisms, of invertebrates, of phyla inherited from a primordial past, of the first cone-bearing trees, of the first flower bearing plants, of mammals, of insects, of trees, and shrubs and flowers. Imagine 86 planets with their own hereditary, evolutionary lines culminating or perhaps on the way to culminating in sentient, intelligent beings with appendages to hold tools, to compose music, to create dance, with tongues to bend around the syllables of languages structured entirely other than any Earthlings can begin imagining. Eighty-six planets with their own dynasties of composers, choreographers, writers, poets, singers of songs. Take all the sounds of all the languages of 86 planets, and all the sounds of all the music of 86 planets, meld them together, imagine the chorus. Now turn down the volume to a whisper: the whisper of the sounds made by the sentient beings of 86 planets. That is only 1/600,000,000th of the sounds of all the neighborhood galaxy's planets, and, of the universe's, a fraction so unfathomable human cognition cannot imagine it.

But this one, this Earth, this Gaia is the one you have. This one, and only this one. Its rocks, its fossils, palimpsests of times more ancient than time, its petroglyphs of a mankind more ancient than language, more ancient than writing. Its horsetails and ginkos, survivors of an unfairytale age of dragons, of cone bearers, of spore bearers, of molds, of microorganisms, of nematodes, of annelids, of the lowliest of beings without which none of our living, none of our songs, or our musics, or our dances, or our writings or our tongues could ever have been possible.

This Gaia is all you have.

2.
Emergency

My neighbor is dying. Once he was on the bus to Mississippi, off to join the Mississippi Summer. He marched with Martin Luther King, Jr.

"Give me a hug," he says to me shortly after learning his prognosis.

"I need a hug, too," I tell him and I hold him tight.

He alternates between moments of great lucidity and hours of night terrors where his scrambled brain allows him at last to collapse into a chair facing the wall, waiting for a reprieve he looks to the dawn to bring. He knows he is dying. He has little time left.

"This is an emergency," he tells me, his still blue eyes wide with terror. "Call the police."

He's too panicked to stay in his apartment. I line up chairs outside along the hallway where I sit with him. After some hours of negotiation, it's clear nothing will do but to call the police. They arrive at 2 am. They ask the usual questions: do you know where you are? who you are? The deep philosophic questions we depend on our finest to formulate.

Judging from the name on her badge, one of these officers is a Latina. "El señor esta muriendo," I whisper. At once the tone changes. The questions yield to appropriate negotiations. The cops promise my neighbor that, come morning, his daughter will return to comfort him.

He summons me for a visit one last time. His eyes are shining. He sits enthroned in his chair in his bathrobe and underwear.

"This is an adventure! I'm going to write a book," he announces, slapping determined hands on his lap. "Here's how it starts:" He gives a sentence or two. "That's Chapter One. Will you edit what I write?"

"Now," he says to me, "now you, Ceil—he always calls me Ceil—you always have things to say about life on the planet; you need to write a book, The Book of Living and Dying. Will you do that?" He waits, all expectation. I've just come to the end of a five-year project but how can I thwart such compelling faith?

"Warren," I say, "I don't know if I can do it, or if I will do it, but if I can do it, and I do do it, I promise I'll dedicate it to you."

He smiles the smile of a mollified child. "Ceil," he says, "this is fascinating!" He means death. Dying is fascinating. "Don't you want to live long enough to see how it all comes out?"

"Dying takes a long time," I tell him, laughing. He has the grace to laugh with me. We are children playing with the red ball of death. We toss it back and forth. "I'm going to write about it," he says. "I starts like this..."

He lies in his old man's bed. More and more he sleeps. Someone comes in to nurse his last hours, to turn him over in the sleep from which he will never awake. Once he marched alongside our great prophet, Martin Luther King, Jr. It ends like this.

Thirteen days later the reactors at Fukushima will explode, scattering deadly fallout over the entire planet.

3.
Move Right Along. Nothing to See Here…

From the day Chernobyl blew, from the time you fell to your knees in horror, you knew this next would come. You know what this means. You remember those hundreds of thousands of young Russian army recruits ordered to the disaster site by the USSR government, who died of radiation agony after building a sarcophagus large enough, and presumably strong enough, to entomb the massive reactor for a few hundred or possibly even thousands of years—but which began to show signs of failure in 1991, a mere five years after it was built.

Around Kyshtym, the site of a Siberian nuclear storage tank explosion in 1957, you know there is a zone where to this day the ground is said to move.

You know what this means. You know the fall-out plume will soon blanket the northern hemisphere. You know it will contaminate the food chain, on land, and on the sea. You know it will taint the soil, and the water, and the air you breathe. You know from now on, it will taint everything you drink and everything you eat.

4.
The Wheel Turns

I want to scatter my goods, explode my possessions, send everything flying to where it may come to rest. Bury it. Bury myself. I feel I have outlived my usefulness. I have outlived my own life. But in the karmic soup I know nothing is lost. In the damped oscillations spanning the world, no effort is wasted. No action, no thought is without eddies, but where in the universe these waves dissipate is unforeseen.

I look about me at the busyness that still goes on around me. My own busyness, my sense of self-importance. From that perspective, I see how this one guards his territory—what he imagines it to be—how that one guards his wealth. How this one gropes for the gematria that will read the future to him.

But in the collective consciousness lives a spec. I know it lives embedded in each living cell, along with the protocol of living; it is the protocol of dying. So in the collective mind—which strives to place it outside—but mind does not live outside. It lies embodied in the living where there is no future and no past. Only one great mind. It carries this message from the beginning, from the first flash of light that dispersed the universes, spun great wheels of stars. All, all carries this protocol: alpha and omega. The beginning and the end.

5.
Being and Not Being

Day five following our planetary catastrophe, I listen to the morning radio reports. After the three reactors, Units 1, 2 and 3, have exploded at Fukushima Daiichi, spewing untold quantities of radionuclides over the Pacific a so-called authority categorically states, "no one will die of this." It's the morning I say to myself: the way I feel, if I didn't have to deliver milk to the homeless shelter, I probably shouldn't leave home at all. Besides it's cold and misting. I slip into my parka.

The sun shines through the branches of a flowering cherry, pink as cake frosting. I'm flying straight into its coppery bark at what feels like 60 miles an hour. My car smashes into the tree with ferocious velocity. The tree buries my windshield in pink, sugary blossoms. It's the last thing I see on earth. Not exactly. My nose bursts. A fountain of red explodes on the curb where I bend, trying to keep the blood from staining my parka.

"You need to lean on something," suggests a tall, black bystander.

"Right now," I tell him, "there's nothing I'd rather do than lean on a black man." He lets me.

A truck driver pulls up, jumps out of his flatbed, passes me a clean white towel, which he urges me to press to my nose. People surround me. They want to help.

I'm not dead. Not yet.

It's five days after Fukushima Daiichi Units 1, 2 and 3 explode, thirteen days after Warren dies, and twelve days after I confide to a friend: I need something to shake me up. It's the morning I wake up to our post-Fukushima world where, just below the waterline, fear comes smashing like a wave that curls above the red stripe that marks consciousness' edge. It's the day two Fukushima workers are found drowned in a basement turbine room.

6.
Floating on a Sea of Drugs

I lie in my bed. If there's pain at all, it's the referred pain of impact, which affects mostly my hips—my Achilles heel. Both have been replaced some seven years ago. These two days seem like an experiment: no mindnoise, no thought, or very little, no anxiety, no self-reproach, no wishing it might be otherwise. Only the recognition that my life is changed. I no longer have a car, will not have a car. Henceforth I will rely on public transport, taxis, a direction I have been moving in for some time, in some small way to defer the moment when the next drop melts off the arctic ice as the life of my planet melts before our unwatching eyes. It will probably cost me the same amount of money. It will only cost me time—a commodity old age generously provides.

The calm I experience now comes close to a state of meditation. I don't wonder if the effect will last. I do not desire it to last, or not. I simply observe. It feels good. It feels marvelous. It feels utterly unlike the way it has felt throughout my harried, driven, anxious, terrified life.

I have met death and it was beautiful. It showered me with a tree's love. It blessed me. It held me fast to this world as its roots hold it fast. Today its bark is scraped, but it still stands straight.

For two days, time seems to move as it has never moved, generously. I have no classes to meet, no appointments to keep, no book in progress. Time stands still, or very close. I like myself in this slow, silent mode, with no brain noise, just breathing in and out, as animals do when they rest. I ask myself: are there perhaps people living on earth who experience existence like this? Is this the baseline, living, living at last, without all of civilization's tortured overlay? Could I have—accidentally—come to some kind of essence?

But the clown mind plays tricks. Two dreams. I fill out a lottery form at the Cal students' union. I win a new car, *un magnifique automobile,* a phrase my mother, daughter of an electric car inventor, never tired of using.

I'm with a student group on our way to SoMa to view a film. It's just past dawn, but I'm speeding through empty, rain-slicked streets. I'll get there ahead of everyone else. In the ticket booth, the woman rubs her eyes. She's been sleeping there all night.

It's a surrealist film I want to see. She reminds me of its name, the one I have forgotten. I feed a bill into the admissions slot, push a red button. A ticket spills its paper tongue out at me. I have three tickets! A whole fistful!

The display case shows stills. A side bar shows a video animation: a woman on a train, a man with a gunshot wound. The woman screams, the image multiplies, the same woman screams again and again, the same man is shot. The color of the theater is wine red, the halls are painted color of coagulated blood, the walls are blood, the floor, the ceiling, the tickets are the color of dried blood—as if I had crawled inside an alien body, but it is not a body I want to inhabit. It is not a film I want to see. And yet, I have sped all my life to get here.

7.
Incorporating the Body That is Earth

I lie in the dark, pondering the enormity, the unfathomable scope of the Fukushima disaster, the degradation to the planet—its soils, its air, its waters. I know the life of the planet has been changed irreversibly. I envision panicked mobs of people seeking the exit doors. But there are no doors. There is no safe ground. There are only time differentials. Perhaps fallout will not hit Bolivia for some time. Who will be the first to wear respirators in the street, or on coming home, to shed street clothing and put it in plastic bags to be decontaminated? Decontamination, we are told, is not a do-it-yourself affair. I see all this visually, viscerally. I think of the seas, the area of the Pacific waters, the size of Texas, the dead zone, choked by plastic.

Once a king made a wish. He whispered his will to the reeds of the river where it entered the gills of the fish; where the eagle caught it up and the bear, and the hunter ate its flesh. "Let me swim in the black gold of the earth, let me have so much I can bathe in it. Let me have enough to drink." Everything his hands touched turned to the black gold underneath the earth. The king died of thirst because that's all he had to drink.

8.
Koyaanisqatsi

The levels of Iodine 131 in one sample of Bay Area rainwater is 18,000% higher than acceptable levels. But there is no problem. In America there is no death. Because in America, there are no headlines. Death is always silenced. We are now living in the after. In a sense the earth has died. But we have not died. It is only a matter of time before all life is extinguished. The earth is a fragile being. She has a finite ability to absorb mankind's progress. Or the conviction of its Great Men that they know what's best: what kinds of energy are cheaper, what kinds of pesticides to use, what genetic engineering will improve crops or animals, what water systems can accept insignificant pollution, what people must be culled, what animals must be shot, clubbed or poisoned. What sea mammals must be deafened, what sonar is vital to national interests, what bombs will "work" best, what enrichment by-product can safely be dropped in the form of depleted uranium (DU) onto the soils of others, who deserves life and who merits death. We Are Become God.

9.
Death Throes

Now in its swan song, once more the human species behaves true to type. Response to a planetary catastrophe? Simple: start another war. Churn up the loudspeakers maximum. Blare the good news: here a bomb, there a mass grave. What fallout? Not here. Not I. Some small, blighted island, contaminated elsewhere. Trot out the humanitarian aid, the smarmy words of sympathy. Spread democracy and birth defects. Bow deeply. Apologize. Above all, apologize.

10.
There is No Story for This

Twenty-three days have passed since the devastation at the Fukushima Daiichi reactors. Plutonium has been discovered in the soils and in the waters of the sea. There are no headlines. Japanese populations have not been evacuated beyond the 20 kilometer (that's 12.5 miles more or less) exclusion zone. The U.S. government advisory urges any Americans within a 50-mile exclusion zone to leave Japan, although for future reference, it should be noted that the U.S. Nuclear Regulatory Commission requires evacuation around all U.S. reactors of only 10 miles. Iodine 131 above acceptable levels has been found in the rainwater as far east as Massachusetts. Americans have made a run on potassium iodide tablets. There have been no official announcements.

I remember my lessons after Chernobyl (I was living in Austria at the time): those parts of Europe where rain had fallen at the time the radioactive cloud passed over became areas of relatively greater fallout. It has been raining heavily over the Bay Area during this time. Days and days of heavy showers. March 24 the storms were particularly torrential. Official sources say nothing of exposure. At this moment, we are told, the EPA is working overtime raising acceptable exposure levels. But suppose there remained a vestige of public responsibility; suppose, for example, there were official bulletins announcing the degree of fallout, suggesting what steps we might take. Potassium iodide pills? HEPA masks? Bottled water? Yes. But there is a time when even bottled water will contain trace elements. A time where everything with which the body comes in contact—by mouth, by breath, by touch, will contain trace particles. If one's house is engulfed by flames, it's best to try escaping to avoid smoke inhalation. But what if one's house is the planet?

Meantime our political Mafia bombs another country, the fourth on the list of bombable countries requiring "humanitarian aid"—a good distraction to keep couch potatoes numbified. The Tooth-Fairy-in-Chief declares, "We're going ahead with plans for four more nuclear plants," and 12 more nuclear subs—this after an earthquake registering 9 on the Richter scale, and several severe aftershocks—and campaign promises to the contrary. In politics, good teeth trump all.

11.
Not Me, Not I

Closer to home, I hear people say, "There's nothing to worry about. The amounts are very small." Or, "It won't happen in our lifetime." (Perhaps because our lifetimes will be shortened by it.) Or "We won't be here when it happens." The important thing to bear in mind is I won't be here. The others—those younger, those not yet born, my children and grandchildren—I won't be here.

"What are you worried about?" my mathematician friend says. "It's a smaller amount than your exposure on a transcontinental flight."

More and more (I answer) I inscribe the planet in my body. Every stream, every rock, every forest, every tree.

"Aha!" she says, "that's why you're so upset."

At last! A language mathematicians understand! But there is no acceptable exposure level. Even more than sudden exposure, steady long-term radiation harms all living things, humans not excepted.

12.
If Only…

The cherry trees offer their froth to the sun. I crouch beneath their cloud of blossoms, crane my neck skyward. Bees! Hundreds of bees dart, crossing paths, near-colliding midair, gorging on pollen—a brief day before the sun, the wind, the rains, scatter petal to ground—a late spring snowfall. I imagine their buzz—if I could hear their buzz—

above the sounds of traffic, car horns, the Doppler of an airplane in flight, the voices of docents touting university life to the young and unsuspecting people of color who have come to attend today's Dream Act Conference. Be Your Own Dream, the come-on says. How long will it take them to discover that the dream has packed its bags all the way to the bank? That when they reach the goal post, they will be ushered into a lifetime of indentured servitude repaying student loans? I scan their innocent faces. I have no words yet to tell them what I know.

13.
Leave-taking

We are striking the theater sets. In the dream we have been up since dawn, packing away the stage machinery, the collapsing aluminum frames, folding them away in footlockers the size of sarcophagi. The great show is over. Done. Yet there are stills, all that's left of a welter of creation, of colors, vibrant colors, a whole museum archive somewhere. It's the work of the theater master who first trained me: a makeup room where actors sit on high bar stools at makeup tables, under the magnifying mirrors and halogen lamps, their shades painted colors of candy: apple to light green, pink, rose, peach. The walls are stenciled in supergraphics. Great bubbles of rose and pink outlined in raspberry against a golden-yellow ground. There's a side exhibit of a stage set: Chinese red, a vanishing perspective of diaphanous red silk panels, screens bearing the slight traces of formal designs.

Now all color has vanished. It's night. I work alone in a Payne's grey room packing away the stage machinery. I am expecting someone. Who? Is it my Master perhaps, come back to stir my mind's unease. But no, it's the Archivist come to ask me if I know where the doors are, the great doors that formed the set.

"Gone." I am quick to brush aside his request. "They're gone." Is it that I am too tired—or too lazy—to find them? Or perhaps that I don't want to be bothered? Where might they have gotten to, I wonder?

14.
Enterprise Zone

My theater master was born in Detroit in 1933. Some of his lightness of being must have come from a place once livened by commerce, music, factories, churches, Saturday night ballrooms, by an American amalgam of neo-classic and mittel-European architecture. Opulence, display, ostentation. A time when a dollar was a dollar, prosperity was just around the corner. Every day in every way, things were getting better and better. They had to: America was clawing its way, hand over hand, out of the Great Depression.

Sixty years later, Detroit rots from within, the inner city, gutted, like New Orleans, like so many other derelict American cities, mill towns, rust belt towns, whole states hollowed out, needle alleys after the great orgies of manufacturing that brought prosperity to the robber barons on the hill. And the people who worked the assembly lines, the dye vats, the looms, who cleaned the hotels, swept the central station, emptied the trash, gone now. Dead of the dying of the poor, sick, unattended. Uninsured, left like their cities to die in ignominy.

This was our shining contribution to the world: convenience, disposability of people, of landscape. There would always be more where that came from. And if labor turned too uppity, there was always Bangladesh, or Honduras. Another Enterprise Zone just waiting to be looted.

15.
Tempest—Without a Body (or Why Walter Benjamin Lives in an Aluminum Factory Now)

Day 29 following the Fukushima Disaster, the MAU Company comes to San Francisco to perform "Tempest: Without a Body." Unlike most performers, they do not represent something, or someone. They are themselves: a company of Kiribati Islanders, some Samoans, New Zealanders and Maoris, some exhibiting the tattoos of their tribes. Although uniquely their own, their idiom bridges Western "high cultural" expectations. The substance of the work cracks established surfaces; it occupies a kind of space along with artists who practiced in less ephemeral forms (forms other than performance): artists such as Goya or Beethoven, or a sarangi master like Sultan Khan whose work is somehow enshrined, not in museums, or recordings, or in concert halls, but in a space that has neither time nor dimension, in human consciousness itself. As I watch the stage events unfold, I have the sense of painting: the use of high contrast, light and dark, the iconic arrangement of objects in space, the emblematic simplicity of gesture—all these elements add up to a quality that touches on the archetype.

An undefined black background where a mottled, metallic surface appears, soon revealing itself to be a monolith. Assaultive sound without any let-up, evocative of a post-industrial and catastrophic world. Blinding overhead lights pick out a ragged female figure with stunted wings. Her each footfall palpates uncertain ground as she walks crouched, bent at the knees, carrying a weight that for all its heaviness is nowhere to be seen. At intervals she pauses, lifts her head, seems to study a distant landscape. Her scream finds its echo in what must be a cavern, vast and limitless in dimension. A static group of male figures appears in semi-darkness, visible only because the skin of their faces reflects a faint glimmer of ambient light. Dressed in black, heads shaved, they slide into the light, toes upturned, sinister automatons on seemingly invisible parabolic tracks. In their divorce from any human quality, they suggest violence and menace.

Bodies inhabit this stage, naked or stripped to the waist, some with the indigo tattoos of the Maori. A torso stripped to the waist falls to the ground to be rolled out of view by a march of headless figures. A body covered in irradiating light writhes in agony as it inches along the length of a light board. The magnified faces of a white colonialist and of a native woman will be projected briefly against the darkness. This Tempest world moves slowly, enveloped in relentless sound, sound that drowns out the desperate plea of the barefoot islander dressed in tie and suit shouting to be heard.

An actor appears walking on all fours, hair tied in knots signifying ears, making fists of his hands, a totem animal. The ragged angel reveals a bloodied hand, and as the blood drips toward her elbow, clouds of red surge over the surface of the monolith. A naked man appears to self immolate in a cloud of smoke. In a scene of final retribution, the stage explodes in a miasma of white light.

The program references twin tower collapse and Klee's angelus novus, the angel of history evoked by Walter Benjamin, but makes no reference to Kiribati's own downwind contamination during the Pacific bomb tests of the 50s.

The performers take their bows. Having shed her dress of rags, the angel reappears with them. She wears a workmanly black shift. It is a sign that what we have witnessed is not the work of actors or dancers. It is the work of people who live in the world, who live in the words they speak, and what they leave unsaid. They are not of our world of simulacra. They inhabit their own reality.

Director/choreographer, Lemi Ponifasio, comes on stage to share the bow, but his noticeably white shirt distances him from the others. And although light is an inextricable dimension of this work, a performer interacting with other performers, the bow is not shared by the only white member of the troupe, Helen Todd, who directs light and tech.

"Tempest—Without a Body" suggests that it's possible to embrace both the mythic roots that make living on earth bearable, and the horror of our time. And that art may be the only resistance left to push back against Capital's final trumpet. In all my years of making theater, of seeing theater, I have never before seen the fusion of such elements in the service of a truly global reach (in planetary, political, historical, and cultural terms). Is it because I have come face to face with consummate artistry? I want to reject that notion as an elitist "Western" concept. I suspect such work could not come into being without its grounding in a consciousness that predates what we like to call history. It is a theater born of tension between the collective indigenous mindset of the "we" and the Western individualist mindset of the "I," yet capable of spanning the contradictions of a world self-destructing in arctic melt, megaquakes, and nuclear catastrophe, a plague world in which only those artists willing to look in its face with eyes unshielded can hope to respond to its realities.

It's the place where Walter Benjamin's Angel of History meets the aluminum factory where many of these Islanders still hold day jobs, survivors all of their once-home island, engulfed now in the rising seas of global warming.

16.
The Angel of History Speaks

The forests turned red. The trees died. When the liquidators came to Chernobyl they began the slow work of killing the earth, of burying the trees and all the things that once lived. They began to enlarge the rent in time's garment. They sprayed the dead trees with glue from the skies to stop the radioactive particles from being carried on the wind. They dug trenches with their great rakes. They buried all the forests with all the red trees with the glue on them until they had turned Prypiat into a moonscape. Bears, foxes, wolves went mad—so it was said. They attacked humans. The humans saw it as madness, but the animals saw otherwise. In the blighted landscape, the humans left one tree standing, crucifix from which, during the war, partisans were hanged. The animals left their bones.

When there will no longer be food on earth, when the bees will have died, we will eat the glass bees of the Blaschkas, father and son, we will crunch the crumbs of glass between our teeth.

17.
Like a Sybil

The liquidators came to Chernobyl with a mandate: leave no stone upon a stone; bury the earth; bury what we have done here. Bury the unholy fire, the fire beyond Prometheus' gift. Bury the trees. Bury the soils. Bury the things people carried in their living: the beds, the cradles, the sheets in which they slept, the toys with which their children played, the sand box in which they filled their pails. The teddy bears, the abortions of women who carried poison in their wombs. Claw away the earth, the trees, the soils, the rivers, the fish. Skirt them in barbed wire. Tear away at time's garment, shred it little by little, rip it away, kilometer by kilometer until at last time itself dies.

Post the warning for future generations in all the languages of the world. Surround the seas with barbed wire and the rivers and the earth so that all may know: stay away. Do not be born here. No life happens here.

— CONDEMNED —
THIS PROPERTY IS CONDEMNED
FOR HALF A MILLION YEARS

18.

I Alone Have Come to Tell You...

What will it sound like, the sound of no time left? Interplanetary winds, explosions of interstellar gases. New universes being born. A wind too fierce for the fox in his burrow, and in the pasture, too fierce for the lamb.

We came to the edge of the cliff. We hesitated at its edge. But those behind pressed forward, all voices of all the generations. Greed claimed the things it wanted, the air, the waters, and the soils. It pushed the frontrunners to the brink. We are falling, falling into the abyss. Falling with all those we have loved and cherished, those whom we have nourished. We are falling in the trenches where our darkest secrets lie buried. We join the bones, time's graveyard of the dead, of the plague, of our wars, of those dead of genocide. We, the living, join the dead. There is nowhere left to go.

The angel alone has come to tell us in the language of the dead. Hair matted, eyes grey, skin grey, she sits hunched in clothes so nondescript she might have pulled them from a free box. She sits between two others who speak for her: the dark voiced woman from northern Russia, the other a former Duma representative. But she, the angel, lives in the world of afterdeath, a messenger from hell. She has returned from hell to tell us. She has a limit of a quarter hour (a quarter hour!) to describe how she put the dead world to sleep. Murmuring, whispering, the words spill out, tumble over her tongue. One third of the earth is poisoned, one third of the air is poisoned, one third of the waters are poisoned. The trees turned red. Whole forests turned to blood. The day was like any other day, the wind was soft, it spread its death as the young men played, the children raced their trikes, toddlers made mud pies in the sand box. In full uniform the military ooompahpahed in the shade of the gazebo; from the podium city councilmen choked on clouds of oratory. People rode the carousel, the tiny bumper cars of the traveling amusement park crashed into one another. Children screamed. Flags snapped in the breeze.

She whispers her sibylance without expression: One third of the forests died, one third of the trees. One third of the fishes, one third of the rivers. As the words rush out, the woman of the dark voice catches a word here, a phrase there. Children were born without eyes, without mouths. Children were born without arms, without legs. The others were ripped from the womb. And of the ones who lived, the cells in their bodies broke in dark masses spreading in their throats.

The soldiers came, the army came. They worked on the roof of the reactor, pushing the debris off the ledge. A human chain. One minute to race up the fire ladders, to sweep, to race down. Up, sweep, down. So many rads, shielded with aprons of lead, one in front, one in back held together with rope. The victims burned in the dark, the slow death of burning. Doctors dared not touch them without catching the slow fire of dying. They worked behind shields, tending their last days, their last hours of agony.

(Next in line to me at the butcher store, a woman said to me, "Russians! It serves them right.")

They sickened and died, whole armies sickened and died, hundreds and thousands of them. Today there remains not a trace. All documents have disappeared, all files have been destroyed. No one knows their names. *I alone have come to tell you.* The words tumble from her tongue in the sibilance of a language no one understands. On the sixth day we looked for something to carry highly irradiated water. We entered the abandoned birthing ward. We discovered milk cans there. Inside we found the fetuses of women who chose to be aborted. I saw their faces, their sleeping eyes, their tiny hands balled into fists. They had turned leathery. The radiation mummified them. We shook them out into the pits. We buried them with the fish, we buried them with the trees, we buried them as we buried all the animals we shot because we could not decontaminate their hair, we could not clean their fur. We buried them in the trenches bulldozed by the great rakes. We buried all the stuff of the living and the dead, the cats, the dogs. We closed dead earth over dead beings. We liquidated the earth and the waters and the skies. We enclosed them with barbed wire.

We hung a warning sign written in letters you can't read, in words you don't understand. This property is condemned for half a million years (Natalia Manzurova, *Hard Duty: A Woman's Experience At Chernobyl,* 28-46).

19.
Fracturing The Earth

In the vast sweep of country from Pennsylvania all the way through Montana, Colorado, Utah, and Texas, giant holes yawn open—poisoned mushrooms in poisoned lands. People with movie cameras travel these roads, filling jars with water. It is water that comes from the taps in the homes of good people, countryfolk. If you hold a match to its flow, the water catches fire.

There's nothing to worry about, the mouthpieces of the fracking companies (BP, Shell, ConocoPhillips) assure us. But everywhere the filmmakers go, they capture the stricken eyes, the dull voices of the countryfolk whose water catches fire when they turn on their kitchen taps, in houses that are no longer habitable. Since 2005, in six short years, thousands of wells, tens of hundreds of truck loads of chemical softeners, of water injected into the earth to bubble the gas upwards. More energy expended collecting energy than the natural gas released can ever provide. A machine, like other machines, machines of people who have become machines. A job.

Who drives the supply trucks? For how many dollars a month? Enough to feed a family. Enough to buy a house where the water catches fire. Running the automaton-powered machine, like other machines: the war machine, the prison machine, the schools machine. This was not ever what we were meant to become. From where in the earth did this poison well spring up?

20.
Out of Africa

Humankind was born one color. It grew and multiplied. It swept into the steppes of Asia, where its eyes narrowed against the sand and wind. Where was the lightning bolt, the small spec of dust perhaps that tripped this freakish pale mutation? Did it trip the mind as well? What poisoned fire did the navigators bring that they could bundle native people in straw, hang them from the rack, hear their death cries as the flames licked upwards? From where did this kind come? From what hole did it crawl into the light of day? Did it serve its apprenticeship by beating its own children?

The messenger alone has come to tell us in a language no one understands. She speaks her syllables to the wind, to the rains, which cannot hear: they buried the reactor, they hid all traces of what happened here. They built a sarcophagus to contain it, of parts put together with other parts. They built the parts at a distance, away from areas too hot for human life. They cobbled the pieces together.

The sarcophagus would last for at least one hundred years.

21.
The Great Men Build Their Houses

In the season of blossoming, the beekeepers come. They set their hives in the apple orchards, in the almond orchards. In the stacks of hives, bees build their comb cities of wax, store their pollen in the hexagonal cells from which they draw to feed their young. Some cells are sealed off now. So as not to feed pollution to their young, the bees close off those pollen cells contaminated by pesticides. One by one, they entomb the cells with the dark wax that shows ashy. These are their buried cells, cells condemned for eternity. But the bees die out. These colonies with their insect Chernobyls are doomed to die. Only the dying is delayed, one season or two. Even bees imagine they can build walls to contain poison.

On their sleeping mats in Japan, people turn in uneasy sleep. They worry their nuclear plants are built on earthquake faults. Their nuclear reprocessing factory is built where three faults meet. Long before Fukushima, their experts warned that Japan has been seismically active for millennia. One of its peninsulas (Shimokita) rose from the bottom of the sea as recently as five thousand years ago. What death wish moves the engineers, the businessmen, the scientists, the politicians to agree to set their fire pits on the chancy seams of the earth's crust?

There is nothing to worry about, the Great Men say. The designs are safe, the plans are safe, designed to withstand a Richter Six. There is no chance of a Seven or an Eight, let alone a Nine. The Great Men have said so. They have decided, and who or what is the earth to contradict?

They have designed the sea walls to rise to thirty feet. The height of a three-storey building. Who is to say the wave of a tsunami might reach to the fourth storey or even a fifth? But the men who know have decided the limit of the water. It will be docile, it will oblige their calculations. It will raise its head only thirty feet. It will humble itself at the Great Men's feet. They are certain of it. There is nothing to worry about.

The Great Men build their houses inland, high in the hills where the great lash of water will not reach.

22.
And Closer to Home, in the Flat Lands

Day fifty-three following the catastrophe at Fukushima Daichi, but all is not lost. In a Berkeley street on an empty doorpost, staplers have been at work:

> My husband planted our buckeye from a seedling he found in Tilden Park, high in the hills. He put it in the ground when our daughter was born. It enjoyed many years in our front yard where it became known as 'the climbing tree.' Then, on October 19, 2003, the worst storm ever to hit the Bay Area took it down. We chipped much of it and put it down as mulch where it joins the earth once more. May it Rest in Peace.

and

> Dear Neighbor: I am elderly and need to walk four miles a day. I rounded your corner and found myself discombobulated: Where's the tree that used to be here? A tree I used to admire every time I passed? Then I read your sign. I want you to have this picture I took of your climbing tree.

or this

> In 1995 a 1,000-mile-long migration of newborn toads, toads no bigger than a fingernail, snaked through Liaoning Province, with adults every 30 feet or so leading the young ones along. Residents of the town of Benxi marveled to see the exodus pass along the Taize River bank.

Why were they migrating in such large numbers—a 1,000-mile-long column of amphibians, each no larger than a fingernail? Where were they headed? What were they fleeing? How could they know where they were going? And how could human observers know the column stretched 1,000 miles? Why did they care? Had they communicated from town to town: Watch Out: The Toads Are Coming!!

23.
Small Holes at the Beginning

All kinds of holes: holes made by indigenous people with a digging stick to plant a seed, holes made by well diggers to tap artesian water sources. Holes made by utility workers to bury pipelines that later explode. Holes made by fracking operations to steam natural gas out of the earth. Engineers now say there's a hole in Fukushima Reactor 1—a force equivalent to destroying the planet. It won't respond to dropping kitty litter in it, not even from an airplane. The planet is vulnerable to a small hole, a hole made in the fabric of time, more and more of them, until time itself exists no more.

It is May 13, two months after the earthquake and tsunami destroyed the nuclear reactors at Fukushima.

24.
Nightmare of the Woman Who Can't Get Out

It's night in one of those god-abandoned working class districts in Long Island, or New Jersey perhaps, where men roam the night and women are absent. Somewhere there's a Heineken's sign flashing red. I sit at a slab table in a near-deserted bar.

I'm here to meet the Patriarch, but there's no talk of ordering, let alone eating anything. He's here to tell me he'll have nothing more to do with his daughter, the youngest of his children, who's homeless, living in a shelter. I get the feeling I know this woman.

What was her name…? Isn't she the one who...?

"Take her back!" I plead with him. "Take her back," I sob. "Please take her back. She's the only one who can save you from this Holocaust."

He leaps up from the table. I race to catch up with him. I find him huddled in a darkened phone booth. I force the door ajar. From what I overhear, he wants to have words with some probation officer, or warden to make sure she can't get out. No matter how I grieve for her, I know he won't let her go.

What part of me is her? What part of her is me? And what part of me is the Patriarch who, like Jehovah, rules his human family with an iron fist?

25.
Tsunami of Another Kind

It's May 20, seventy days following the nuclear catastrophe at Fukushima Daiichi. The State Department issues a secret letter to Japan guaranteeing there will be no effort made to bar its foodstuffs from entering the United States. The levels of Cesium and Iodine 131 are shown to be exceptionally high over the northern hemisphere, over continental United States. Japan has asked the United States to cooperate in not releasing fallout information. The EPA no longer reports the levels of contamination sweeping over the continent. The Norsk Institute web site, where these findings originally figured, has been taken down. But there are numerous mirror sites where day by day anyone can watch the radioactive clouds swirling over the waters and the land.

A bill is before Congress allowing government shutdown of the internet.

An authorization before Congress declares that henceforth the president will be able to declare war throughout the world, anywhere, at any time, bypassing even token congressional approval.

According to the most recent judicial determination, police are now entitled to enter homes unannounced and warrantless; resistance will henceforth be criminalized.

In the heartland, the FBI questions children without their parents' consent.

The administration uses the states secrets privilege to hold itself immune from prosecution by victims of its policy of "extraordinary rendition," a butcher's name for torture.

In the country of death, the collapse of the Planters' Revolution is nearly total. Just a few more steps before the dark. This, too, is a tsunami, but of another sort. Unless…unless…

26.
The Mouth of Fear

Sometimes, in moments of perplexity, I draw on the tarot. Strength, the eighth in the sequence, shows a woman prying open the lion's mouth, or does it show her closing it?

If she forces it shut, is she under the spell of her own fear? Does the lion swallow her? If she pries open the lion's mouth, has her voice become one with the lion's voice?

Is the woman inside the lion? Or is the lion inside the woman? Does the woman speak? Or does she stay silent, locked in the phone booth talking to the warden?

It is seventy-one days now following the catastrophe at Fukushima Daiichi. I feel restless, wake up from dreaming, afraid I will not be able to write my dream exactly as it happened. I force myself to record it before it vanishes:

I am standing by a river bank where the shallows cause the stones to chatter with the churning of the waters. Through the chill morning mist, a fur-bearing animal picks its way over the stones, grappling for a foothold. A fox? Too large, its coat too thick and tawny. A wolf crashes through the underbrush. Clamps my hand between its jaws. I feel the ridges of its mouth, the wet of its tongue against my palm. I still my fear. The wolf frees my hand at last. I run my fingers through the ruff at its neck.

"There," I say, as if something has been settled.

The Chauvet Cave, where cave artists drew images of animals some 30,000 years ago, crouches at the margins of the Ardeche River Gorge.

Their drawings are the subject of Werner Herzog's 3-D documentary "Cave of Dreams," where the roving lights of the filmmakers animate herds of animal paintings, which predate those of Lascaux by some 20,000 years. Bears, leopards and lions stalk the tunnels; and here a flock of aurochs races, pursued by wolves. A stampede of horses wheels on the curved surface of the stone. A bison with human hands nuzzles a woman's sex. In the dust bear tracks come alive. At last the lights come to rest on the small footprints of a boy alongside the footprints of a wolf.

An anthropologist remarks: "Ancient men may have thought differently; may have seen their world differently than we do. It was a fluid world where a man could become an animal, or an animal could change into a man. It was a permeable world, where a man—or an animal—might communicate with a spirit"— one who may have lived some years in the past, or 30,000 years into the future.

27.
Preserving the Moment

The landscape of the Ardeche was not as it appears today. Thirty thousand years ago it lay buried under 3,000 feet of snow. To keep warm, day and night, the people who lived there wore reindeer parkas, boots of hide. If they wanted to eat, they had to hunt their prey, which often hunted them. They huddled in clothes of their own making, built their own fires. If they wanted to illuminate their caves, they did so by torchlight. They made their drums from the skins of animals and their flutes of human bone. They speared animals: aurochs (the ancestor—now extinct—of today's domesticated cattle); bear, cave lions, deer, bison and rhinoceros. They lit fires, sheltered in caves, cooked raw meat over fire pits fueled by wood. Their way of life remained relatively unchanged for some twenty-five thousand years.

Some 20 or so kilometers downriver from the Chauvet Cave, cooled by the waters of the Rhone, the Cruas Meysse Nuclear Power station went on-line in 1983. It sports an ecologically-inspired mural depicting water and air that required a crew of nine mountaineers and 1,000 gallons of paint. Cruas Meysse heats (and cools) people's homes and swimming pools; it illuminates their nights; it allows them to cook at regulated temperatures. It supplies between 4% and 5% of France's energy, but it may not be in its present location thirty thousand years from now. Like the other 52 of France's 59 power stations, it is located in a seismically active zone.

To preserve its rarest of cave art from the molds that result from the breath of curious crowds, the Chauvet Cave will be forever closed to tourists, but plans are already underway to create an amusement park nearby, where for the next 30,000 years, tourists can pay to see what planners promise will be its exact replica.

28.

This is Our World: Far From the Golden Temples

It's the seventy-second day following the Fukushima Daiichi nuclear disaster. The world gyrates in a Dance of Death. Art historians are familiar with this early 14th century iconography. Its most famous example is by Orcagna and, until U.S. bombs melted its lead roof, it graced the walls of Pisa's Campo Santo. *The Seventh Seal,* Bergman's iconic film, ends with a shot in which the Knight and his household are depicted in silhouette dancing over the distant hills toward their certain death—a very specific one—of plague. But plague rolled up its banners before our living memories. Most people alive today don't remember the ever-present threat of polio; most don't yet suffer from the ravages of AIDS. In the heart of the shopping districts in the cities of the West, there are no traces of death. The dance has become a feeding frenzy of fashion-conscious shopping, shoe fetishism, self-encapsulation in cars, by iPod, by cell phone. The breakdown of the collective sense is all but complete. Only strangers older than the threshold age of 60 seem to remember how to talk to other strangers.

Far from the glitter of commercial zones, in the alleyways beyond the chrome and shining lights, another law applies. The poor—mostly black folk—haunt the doorways of abandoned buildings, huddling against the chill.

It's a short walk from the raucous café where a band of musicians promotes Friday-night jazz. I can't help thinking they'd benefit from better acquaintance with the blues—with people who sing and play informed by pain. But most of these folks don't seem to share much of that history, or if they do, it seems not yet to have made its way to their vocal cords or fingertips.

Outside, the street is near-deserted. It's a four-block walk to the bus stop in the chill night air. In the shelter of a doorway, someone stops me. "Ma'am, you got something for an old man?"

I study him. His face is marked by living—what? Homelessness? Joblessness? Backbreaking work? Jail? His face tells all these stories. I pass him a bill. He thanks me. Now it's up to me. "Thank you for asking me," I say.

The night—although it's late May—is bone-chilling. An ancient black woman waits in the shadows by the bus stop, huddling inside a fleece blanket. She keeps a tight grip on a black garbage bag propped atop an enormous wheeled suitcase. The bus pulls up. As I thread my dollar through the cash box, people pile in behind me. The ancient woman still stands at the curb. The driver is about to close the door.

"Why won't you let her in?" I ask him. "She's an old lady. She's got lots of stuff."

He whispers, "There's a camera. I can't take her in for free. I'll lose my job. I already took one person in for free. I can't take another one."

"You should have asked. I'll pay for her," I say. "Come on, come on in," I say to her. "It's cold outside."

She takes her seat. I wonder where the darkest hour of the night will find her. Who will take her the next step of the way? And the next one after that? Who will be there for her? And is her life—and mine—made up of these small things that linked together spin the thin and fragile thread that keeps both of us human?

29.
Every Man for Himself

I sleep intermittently through the night. I lie awake thinking of my life, the ties that hold me dear to a world that more and more seems to slough its people off. I think about myself in the web of my few friendships, among my many alienations. I think of the man from Burundi I meet in the street who tells me he hasn't eaten since yesterday. "Why are you Americans so unhappy?" he asks. I think of my neighbor who spends days isolated. I wonder what he does and where he stays. Does he think about himself? Does he call himself depressive? It's going on four years we live in close proximity. Four years, I tell him, and still I know nothing about you. To me you are a cipher.

"I have wasted a decade," my cipher-neighbor says. He's talking about the moment-to-moment struggle to keep present to one's own humanity despite the poignancy of knowing that, although the reach of mind is governed only by the limits of imagination, the body's prison is mortal, and, sooner or later, will announce its small betrayals.

Like the personal body, the earthbody, too, must be prey to its own betrayals: tornadoes, floods, fires, hurricanes, landslides, megaquakes. I revisit the mist of dream. I watch the wolf retrace its way over the chattering stones, crashing through the underbrush to clamp my hand inside its jaws. I feel the wet of wolftongue against my palm, the ridges of its hard pallet: passage through the ridged and wet gallery that permits emergence into life. An opening, narrow, the sickeningly narrow passage like the claustrophobic galleries of the Chauvet Cave, where we first crawl on our knees, handhold over handhold, pressing forward in the dark to emerge at last in the chamber imaged with herds of stampeding animals. But without torchlight, in this cave we are born blind. And with torchlight, the cruel limits of the cave—the prison of our prison—become evident.

It is May 23, the seventy-third day following the Fukushima Daiichi planetary disaster. Eleven years since 2000, when with one vote, the high court assured the final collapse of the Planter's Revolution. That one lives in a dictatorship is not sufficient reason to declare one's death; no reason to embrace silence or invisibility. In its deepest marrow my dream body makes its choice to live.

30.
What Would You Do?

The seventy-third day since the earthquake and tsunami ravaged the nuclear installation at Fukushima Daiichi the Japanese government and TEPCO, the Tokyo Electric Power Company, announce that in addition to the meltdown in Reactor 1, there are meltdowns in Reactors 2 and 3. (Even more dangerous, spent fuel is stored in Reactors 4 and 6, although the reactors themselves were shut down for refueling at the time of the disaster.)

"If you had three grandchildren, what would you do?" It is the first right question I hear spoken since March 11—and from a stranger—about the destruction of our world. It is the seventy-third day, the day TEPCO announces that there are meltdowns in Units 1, 2 and 3.

31.

The Whole Ball of Wax

Dr. Christopher Busby from the European Committee on Radiation Risks is interviewed. He says the events now occurring at Fukushima are not regional but global. He says they are out of control and that TEPCO and the Japanese government don't know how to deal with what is happening.

But Bechtel wants to lend its expertise. Aware that Japan desperately needs a pump of its manufacture, Bechtel ups the slightly more than $700,000 asking price to a little under $10 million dollars. It is the seventy-sixth day following the disaster at Fukushima Daiichi.

32.
Botanical Inventory

Today marks the 78th day of collapse at Fukushima. Astrophysicists tell us that somewhere in a narrow tranche of our neighborhood galaxy, there are 1,235 planets of the appropriate size, slightly smaller, or slightly larger than Earth, although of these only 86 are in a favorable relation to their suns, neither too far, nor too near, to make the presence of water—and of life—possible on their earths. Are they telling us something? Why now? What do they know that we don't know?

On my own Planet, XLB365, there is no American dogwood. The dogwood on my planet is different. Watered by XLB365's spring rains, prodded by light beamed on it by XLB365's rotational tilt, reminding the roots that warmth is just around the equinoctal corner, it sprouts rows of tiny white, embryonic flowers, springing like candles among branches tiered like pastry trays. On the highest branches, flower clusters blush a bashful pink.

On XLB365 my jack in the box grows three feet tall. Its dun-colored stem is mottled like a lizard's hide, its texture smooth and cool as a woman's skin. Its throat is pale and moonlike as is its spadix. Its hood is the color of ox blood, its spathe the size of a hearing trumpet; it throws out a four-inch-long hair-trigger tongue, a thread-thin causeway for small flies to stop on their rounds just long enough to pollinate its heart.

Across a narrow inland sea grows a smaller flower whose spathe blooms a moony apple green, the color of lettuce that's been left in vinegar too long. Different languages are spoken on XLB365, depending on where native speakers live in relation to these flowers and the colors of their spathes.

More information on XLB236, in what tranche of the Milky Way my planet can be found, including its celestial coordinates (and instructions how to get there) will shortly be forthcoming.

But for now, advance notice from astrophysicists remains primarily mathematical in nature, revealing a certain innocence of mind. They define as XLB only those 86 Excellent Bodies, which they identify as containing the possibility of water. They make no mention of dirt—the stuff that makes things grow.

33.
Learning From Past Mistakes (or Is Intelligent Life Possible Following Planetary Collapse?)

Fukushima Daiichi foreshadows more nuclear events of similar magnitude not only in Japan. With the melting of the polar ice, we are told Earth's loads are being redistributed, triggering increased seismic activity world over, particularly around the Pacific Plate (or Ring of Fire as its rather evocative name has it), where on California's West Coast the Diablo Canyon nuclear power plant and San Onofre are located—directly over earthquake faults.

Tectonic Plates

If the horse's mouth says it better, and with more authority, it may be time to take dictation. In *Nuclear Power Plants for Tokyo,* Takashi Hirose takes the logic of the nuke promoters to its absurd conclusion: "if you are so sure that they're safe, why not build them in the center of the city, instead of hundreds of miles away where you lose half the electricity in the wires?"

He says that nuclear catastrophes such as those happening at Fukushima Daiichi may signal the end of our world. Because the Pacific Plate, the largest of the plates that envelop the earth, is in motion, there will be major earthquakes all over the planet. In support of his argument, he cites the September and October 2009 earthquakes off Samoa, Sumatra, and Vanuatu, of magnitudes between 7.6 and 8.2, each located at the boundary of either the Pacific Plate or a plate under its influence, followed in January 2010 by the Haiti Earthquake, at the boundary of the Caribbean Plate, pushed by the Pacific and Coco Plates, and in February by the huge 8.8 magnitude earthquake offshore from Chile.

Hamaoka

More locally, Hirose points out the Japanese archipelago is directly over the place where the Philippine Sea Plate, the huge

Pacific Plate, the North American Plate, and the Eurasian Plate all meet and at the very center of the area where these four plates press together is the city of Omaezaki where the Hamaoka Nuclear Power Plant is located. Sonar readings at the site indicate that from thirty years back, the Eurasian plate has been bending, which means that it is in a condition where it can be expected eventually to spring back. Hamaoka is the only installation—so far—shut down by the Japanese government because, according to Hirose it is located directly above where the next major earthquake is predicted to occur.

Death Ash

Even more troubling than nuclear plants are re-processing plants. The Rokkasho plant, where the nuclear waste (death ash) from all the nuclear plants in Japan is collected, where the plutonium, the uranium, and the remaining highly radioactive liquid waste is separated out and where 240 cubic meters of radioactive liquid waste are now stored, is located on land under which the Pacific Plate and the North American Plate meet, plates which are now in motion deep beneath Japan. Rokkasho is built to withstand only 20 to 25% of the magnitude of the quakes rocking Japan today. The Shimokita Peninsula where it is located is an extremely fragile geologic formation that lay at the bottom of the sea as recently as 5000 years ago; if an earthquake occurs there, Rokkasho can be completely destroyed. To quote Hirose, Rokkasho "is the most dangerous factory in the world" because its liquid waste constantly generates heat and needs to be continually cooled. In an earthquake, any damage to pipes or electric flow interruption could cause an explosion which would expose persons within a 100-kilometer radius to radiation 10 to 100 times the lethal level, presumably meaning instant death.

According to Dr. Hirose:

> On April 7, just one month after the March 11 earthquake in northeastern Japan, there was a large aftershock. At the Rokkasho Reprocessing Plant the electricity was shut off. The pool containing nuclear fuel

and the radioactive liquid waste was (barely) cooled down by the emergency generators, meaning that Japan was brought to the brink of extinction. But the Japanese media, as usual, paid this almost no notice.

So far in the aftermath of Fukushima Daiichi, the United States Government has shut down exactly none of its total of 104 plants, at least 6 of which are in highly seismically active zones, imperiling millions of people, and 23 of which, like Fukushima's, are GE Mark I boiling water reactors. El Presidente guarantees government underwriting of four more nuclear plants; and the U.S. managed media is reporting exactly nothing concerning current levels of fallout over the United States.

34.
Little Absurdist Theater Piece...

"Onkalo" is the subject of the film by Michael Madsen describing Finland's plan to bury its radioactive waste. Onkalo is the site—like Yucca Mountain—where the Finnish Atomic Energy Commission has chosen to bury nuclear waste for the duration of its afterlife—a skimpy 100,000 years, although the half-life of one radioactive element—Uranium-238—is 4 and a half billion (that's 4,500,000,000) years. Eleven men, and one woman sit at a table, talking, speaking into the camera explaining Onkalo:

Onkalo? Onkalo is a place where we have buried something—something from you—to protect you.
And who is this you?
Well, not you exactly. Your descendants. You won't be around 100,000 years from now.
Protect my descendants from what?
All of Finland's nuclear waste.
And what about the waste that Finland continues to generate once Onkalo is filled?
We will dig more Onkalos.

And how will you make sure my descendants won't find them, won't yield to their natural curiosity to uncover your strange remains?
Not a problem. We will label it Keep Out.
In what language?
In every one of 156 UN-recognized languages.
No language to our knowledge has ever survived 100,000 years.
No problem. We will leave a symbol.
What kind of symbol?
A universal symbol, one anyone can recognize.
For instance…?
Some kind of universal sign. A kind of image. An image of a forest with nothing in it but thorns.
But what if the earth's climate has shifted. What if there are no flora displaying anything like thorns?
We will erect a kiosk with an archive.
And archive of what?
Culturally sensitive signs that all will understand.

Madsen calls "Into Eternity" a film for the future. Its opening words set the agenda:

> *This is a place where we have buried something from you to protect you. And we have taken great pains to insure that you will be protected for 100,000 years, and we want you to know that this is not a place for you to live in. Then you will be safe from this place.*

But nothing man-made has lasted more than 1/3 of 100,000 years. Scheduled to be completed more than a hundred years from now, Onkalo will be the longest lasting sign of our civilization. What will it tell you about who we were?

By contrast, the designers of Yucca mountain don't mince words. No kiosks, no warning icons, just the statement preening with the self-delusion of an empire in decay: "It seemed to us that we were a VERY GREAT PEOPLE!" (D'Agata. *About a Mountain,* 127).

35.

Body Without an Anus

The magnitude of the problem of storing spent fuel (death ash by another name), is one scientists, among them Albert Einstein, Lou Szilard, Neils Bohr, Robert Oppenheimer, Edward Teller and others left for last. Had these geniuses, many of them self-identified, allowed their dangerously tunnel vision to wander sideways into the biological sciences, they might have noticed that all living systems, no matter where they might place in any hierarchy, on this planet at least, are organized to take in nourishment, use it to generate energy, and excrete by-products, which are of biochemical benefit to other planet-dwelling organisms in the shape of food or fertilizer. Nowhere in this recycling system is there provision or place for toxogenic organisms, only those in the service of the cycle of life. But the great physicists overlooked the obvious in their mighty pursuit. The question of compatibility in a biogenic context never entered their minds. They designed a non-recycling system—a system without an anus—in a re-cycling world.

The problem the Onkalo "experts" address has to do with the billions of years decay period of radioactive waste—although they limit their provision to 100,000 years. Is it better to try to hide the existence of Onkalo, to conceal mankind's dark secret below ground, in solid bedrock for 100,000 years, or to thwart humankind's curiosity by designing a mythology robust enough to warn the generations of humankind to stay away—at least for the next 100,000 years? And what would such a mythology look like? And how could it survive in the face of unstable conditions on the surface of the earth, such as natural or manmade disasters like wars, or flooding, not to mention the cultural and linguistic problems related to the passage of time? Mankind has never before created a project of such cosmological duration. Our most enduring of such projects, the pyramids of Egypt, spanned several dynasties, and, collectively, took 2,700 years to build.

Onkalo asks for nearly 40 times that stretch of time. It presupposes that humankind—in some form at least—will be found on earth 100,000 years from now; it presumes that in the context of some future culture, humankind may still be able instantly to recognize the warning symbols these "experts" dream up, images such as blasted landscapes studded by thorns, or possibly with a warning underground station that, even in its projected contemporary incarnation, looks suspiciously like a religious shrine, and it assumes that the possible deterrence value of these symbols will somehow deflect surviving humankind's natural compulsion to explain the unexplainable and to unearth relics and writings of all previously vanished civilizations.

And here they are, a *grand guignol* of experts, all neatly lined up for our inspection, talking, speculating, warning, advising, describing, explaining, justifying, rationalizing, unable to wipe the smirk of satisfaction from their expert faces. Admitting—finally—that nothing much is certain. Admitting—finally—that even they don't really know.

36.
Theatrical Security

If the Swedo-Finnish attempt to bury the traces reads like black comedy, not surprisingly, the U.S. version qualifies as farce. Politicians raised on Kansas corn, wide blue skies, and TVacuity delegate themselves to making decisions and appointments better left to people who claim at least a smattering of expertise in such areas as geology, ecology, not to mention nuclear physics.

In the U.S., three presumably appropriate sites were chosen to store nuclear waste. By default, because Nevada couldn't muster enough NIMBY (notinmybackyard) votes, the choice fell to Yucca Mountain. Massive PR campaigns were mounted to convince the public that nuclear was grand, and nuclear waste was even grander. Throughout Nevada educational sites were stocked with displays and teachers' manuals @ $1,000 bucks a pop. Billions of dollars were allocated, besides digging, to equipping underground test laboratories to determine how best to store the waste. And in what to store the waste. How long to store the waste seems not to have entered much into the discussion. Although the half-life of Uranium-238—only one element of radioactive waste—happens to be 4.5 billion years, 10,000 years was chosen because the EPA must have thought it looked pretty good and, like Homer Simpson might have said, 10,000 years is a real long time.

Bob Fri, chairman of the Board on Radioactive Waste Management, when asked the reason a 10,000 year time scale was set, shrugs. Well, yes it was, maybe yes, and maybe no, because "basically Yucca Mountain is what happened." Originally Fri and his colleagues suggested one million years might be a better place to start. But somewhere, because the stability of Yucca Mountain was in question, government policy wonks got busy and reduced the period 99%.

Hired by the State of Nevada after having successfully defended other states as proposed sites for waste depositories, Bob Halstead describes the Yucca Mountain Project as "an exercise in planning for a nuclear catastrophe that is fundamentally rhetorical. It's theatrical security because the preparations that are being made by the Department of Energy have no real chance of succeeding. They satisfy the public...because they're a symbol of control ....This waste is going to be deadly for tens of millions of years....[But] the biggest threat we face is the transportation of this shit" (D'Agata. *About a Mountain,* 68).

According to D'Agata's dizzying mathematics, transporting the 77,000 tons of spent fuel from the pools in which it is presently stored throughout the U.S. would require 3,000 truck deliveries per year for 40 years, or one load every 2 hours and 48 minutes, converging in Las Vegas traffic at the intersection of Interstates 15 and 80 in a stretch of interchanges so confusing locals refer to it as the "spaghetti bowl."

Radioactive Waste Management's worse case scenario describes an accident in which one such truck rolls over, catching fire in rush hour creating a nightmare commuter backup plausibly involving 10,000 cars piled behind a truck on its way to Yucca Mountain that has rolled over and presumably caught fire. Diesel trucks become engulfed when their core temperature reaches 1,832 F degrees, but nuclear waste casks become engulfed at 1,732 F degrees, splitting open a half hour later, and exploding their particulates into respirable aerosols, allowing not even a half hour to evacuate all the piled up cars, and all the drivers, and all the passengers of all the piled up cars backed up along the freeway. When the firefighters arrive, their hoses are not long enough by 300 feet to reach the site of the accident.

According to a Sandia National Labs report such a "clean up would not be possible with today's present technology....The only two possibilities...are to raze and rebuild the site, or to [declare] the area uninhabitable." As D'Agata points out, "When it comes to a place like the City of Las Vegas, [the DOE's] estimate of 1-in-10 million odds over a 40 year period [collapses] to 1-in-27,000..., making the possibility of a nuclear accident in Vegas higher than the possibility of striking it rich in a casino."

Yucca Mountain was discovered to stretch and move, its mass composed of 10% water. To determine the safety of storing metal waste disposal casks in its vaults, 63,000 gallons of water were poured over its summit. Three months later, 63,000 gallons had collected in the underground vaults. To waterproof the facility, DOE proposed a miracle alloy, Alloy-22, which was designed as an impregnable defense against water infiltration and other geologic events. Challenged by a congressional review board, the DOE staged what may have been the all-time Marx Brothers press event. Ceremoniously, two lab-coated geoscientists placed three samples of Alloy-22 in separate glass beakers to which they added a mixture of water containing the same minerals found inside Yucca Mountain. Cameras rolled. The press took notes. The public held its breath and waited. At a signal, the scientists removed the miracle alloy from the beakers. "The cameras zoomed in....The metal had corroded in 21 minutes" (*About a Mountain,* 57).

Twenty years and 30 billion dollars (that's $30,000,000,000) later, the Yucca Mountain project has faded from El Presidente's 2012 budget. The project's management office within the DOE was closed last year, and according to the Office of Management and Budget, activities on nuclear waste management "are now being performed elsewhere." Such hyperbolic theater might have been avoided had the U.S. Government and its agencies listened to the Western Shoshone, whose sacred land this is, and who knew long ago that Yucca Mountain had a way of slowly swimming westward.

37.

Nimby: Somewhere Else, but Not in My Backyard...

I still hold onto the much abused book of *Tales from Other Times* my father must have picked up some time in my childhood from a Fourth Avenue bookseller for 55 cents. I remember the words of "Little Brother, Little Sister:" All the waters from all the springs were poisoned. When little brother went to drink, the poisoned water turned him into a fawn. Little brother never returned to his human origin; he never regained his place in the human gene pool.

In Kosovo, in Iraq, in Libya, the United States and its dark proxy, NATO, have made use of depleted uranium ordnance. The rationale offered is that it is capable of piercing tank armor, but the unstated reason is that it is one way of getting rid of all the U238, the by-product created in the U235 enrichment process—but not in "our" backyard. All the same, returning troops from various theaters of war bring it home—to the United States (and other NATO countries), in the form of radiation sickness (otherwise known as Gulf Disease), cancers, a high incidence of leukemia and birth defects among their offspring.

38.

Inventory Before the Fall

Despite S.F. MOMAs hyperventilating PR, "The Steins Collect" is an overrated exhibition, all except for the "Portrait of a (green) Woman" (who on the day Matisse painted her was actually wearing black), an insignificant scrap of canvas, thrown together in haste, a casting of the dice in ways never seen before. This was a trace of what was possible on this planet in 1905, a reassembling of nature in utterly astonishing ways. When exhibited in the Salon d'Automne, other painters slapped their thighs, bowled over in derisive mockery, unable in the moment to make room in their carefully preconceived firmament to welcome a painting that celebrates light itself with every improbable juxtaposition of color, one of the first times human imagination allowed itself to fracture rainbows.

Where I live, on the fifth floor of a concrete, 1964, pier slab constructed building located approximately one-half mile from the Hayward Fault the following artifacts occupy the walls and some other surfaces:

Da Vinci: "Madonna, Christ Child and St. Anne," 16th century, Masolino da Panicale: "Annunciation," 14th century, Unknown Spanish master, "Nativity," 10th century, *Mali ceremonial mask, 20th century, *Japanese woodblock print, early 19th century, *Two pages from sacred Hindu codex, 19th century, *Indian wall painting, "Krishna & Gopis," 19th century, *Japanese koto, *Indian bronze wheeled horse, *Indian bronze temple lamp, *Indian ninth century head of the Buddha, *Lurçat needlepoint tapestry, *Page from Thai Ramakien, *Rajastani miniature, on parchment, 19th century, Persian 18th century illustration, *Indian mughal miniature, *Persian miniature from the Sha Nama, *Indonesian Batik, *2 Japanese 18th century tansus, *two etchings of Auschwitz, Venetian 17th Century engraving, *Rajastani 19th century mirror appliqué wall hanging, and—full circle—*Balaji, blind hermaphrodite god/goddess of ordinary people, purple heartwood carving, Indian, 19th century (asterisks indicate originals).

The 140-year periodicity of quake occurrence on the Hayward Fault suggests that a major one is overdue. In the era of major quakes around the Pacific Plate, we expect the magnitude of the next one to hover at plus/or/minus Richter 8. In such a quake, the building will most certainly be heavily damaged if not destroyed. My own life may end, and my artifacts, the oldest of which only dates back to the ninth century of the "Christian" era, will be damaged if not entirely destroyed. Their destruction will subtract their small and modest number from the immense inventory of visual art that has been created during humankind's short passage on this, our planet.

Consider the magnitude of the entire earth's museum, take all the traces of artistic creation since humankind first made cave art at Chauvet some 30,000 years ago: the works of DaVinci, and Monet, the temples of Hampi and Chartres, all of it, and now: take it out into the streets of all the cities of the world; slash the canvasses, smash the carvings, pulverize the sculptures, explode all the temples and cathedrals. Leave not a single one recognizable, no stone on stone.

Even so, that destruction is only a fraction of loss measured against the total loss of Earth.

Consider now decommissioning those nuclear plants worldwide that happen to be located above earthquake faults. Now consider decommissioning any nuclear power plant that happens to generate nuclear waste with a half-life of more than 100,000 years. Is such a program possible on earth? Is such a program likely? Will such a program be realized in time to prevent the next Fukushima and the one after that? What does the end of the world look like to you?

39.
No See'ums

On the ninety-third day following the planetary catastrophe at Fukushima Daiichi, the Tokyo Electric Power Company (TEPCO) admits that not only are there three meltdowns, but also the degree of contamination is many times more severe than initially reported. And although the soils inside school playgrounds have been decontaminated, outside of school grounds, 30,000 children are being exposed to degrees of contamination, which will result in serious health consequences to them later on, but they are being issued toy-sized Geiger counters to protect them.

The poignancy of the human condition is that we are born knowing—or soon to discover—that we must die. Life's steeplechase runs us over water obstacles, high jumps, and ultimately we learn—most of us—our verdict: guilty of life. In old age—most of us—receive the sentence: cancer, coronary insufficiency, aneurism, etc. without possibility of parole.

The planet, too, is guilty. Guilty of harboring the human species, the only biomorph without natural predators. That fact, boredom, and year-round estris have allowed humankind uncontrolled proliferation, and with it has come an insatiable need for energy, wood fires, coal, oil, natural gas, and finally, nuclear energy—our sentence without possibility of parole.

For a magpie civilization that increasingly enshrines the shining surface of things, what better way of death: we can't see it, we can't taste it; we can't hear or feel it. Death has been conquered at last: it has become fully absent.

Like global warming (climate collapse by another name). As long as snow falls—lots of it—on the center of the earth, in Washington, D.C., there is no global warming. As long as the Potomac doesn't flood, no Katrina. As long as forest fires don't incinerate Rock Creek Park, no Russian or Arizona fires, as long as caviar and champagne keep getting flown in, no famine, and no drought anywhere, not even in Somalia where right now, two million people are at risk of starvation. Nuclear energy (at Indian Point, New York, and Vermont Yankee) is sufficiently distant to be perfectly safe and even though you can't see its byproducts, they're clean, especially at that distance! They're even good to eat.

In the 20 km Fukushima exclusion zone, cows fed on straw containing cesium fifty-six times the allowable level were slaughtered. Their meat was distributed to several prefectures. An entire cow's worth of meat was consumed in one radiation-hungry restaurant alone. Some of this meat even made its way into school lunches. Cesium-137 has a 30-year half-life; cesium-134, two years. If cesium enters the body it may spread to the muscle and other organs and cause cancer. But according to Goshi Hosono, "eating only a portion will not cause a great deal of change to one's health." He should know. Hisono is State Minister in Charge of Food Safety in Japan.

On the eighty-sixth day following the nuclear catastrophe at Fukushima Daiichi, Greenpeace announces that marine life within a 50-mile radius off the coast of northern Japan shows levels of radioactivity well beyond "acceptable" limits.

But repeated evidence shows there are no acceptable limits for living beings. Dr. Alice Stewart's research makes the point. First to discover that X-rays caused heightened rates of cancer in fetuses photographed *in utero,* she with her colleagues, Dr. Mancuso and statistician George Keith reviewed the findings of Hiroshima/Nagaski survivors, people living downwind from Windscale and nuclear workers exposed over long periods to low grade radiation at Hanford to discover that all groups show heightened rates of cancers, leukemias, infant mortality and birth defects (Gayle Greene: *The Woman Who Knew Too Much: Alice Stewart and the Secrets of Radiation*).

40.

Tunneling: Chauvet, 28,000 B.C.—Onkalo, 100,000 A.D.

Because humankind is first said to have walked the earth some two hundred thousand years ago, is it a foregone conclusion that we will still leave footprints somewhere around here 100,000 years from now? Has our long way taken us from cave to cave?

On the walls of Chauvet cave splash print of hands, here, there, marking what? Our incantation, our seal? Our kilroywashere? Our flocks of hands, our flocks of deer and elk and wild boar. Wild buffalo, burned in rock face, accented with the occasional gash of red.

Will this first evidence of art be replaced 130,000 years later by techno-art: canisters of radioactive waste, sealed vessels, rusted, corroded, buried under stalagmites and stalactites of sugary, sparkling crystal? Can we imagine this, mankind's journey, no different from that of earthworms, from tunnel to tunnel without possibility of light? When all this time our history might not have been marked by centuries of labor under earth, drilling for coal, dying of mine explosions, hardened lungs, blackened by coal dust; blasting of mountains, flattening whole landscapes, dumping the tailings in streams—streams once so clean people could cup their hands and drink—blasting poisons into the water table until the water itself catches fire, spreading invisible waste over land where people of another race and belief must raise young born with damaged limbs or brains, or eyes, or ears because there's no other way to store it? Turning whole landscapes to uninhabitable zones, where even humankind's best effort—Onkalo—is massive, cruel self-deception? When all along—irony of ironies—the same light that awakened our sight, our gift of color, could have cooked our food, lit our nights, turned our wheels, warmed our winters, dispelled our darkness? When 100 years after James Watt perfected his steam engine, Augustin Mouchot displayed a solar-powered steam engine driven by a parabolic mirror at the Paris World Exhibition of 1878!

Is this the last chapter? Is this how the end looks like to you? Does it scare you? Scare your children and your grandchildren if you're cursed enough to have them? What are your regrets? Write them here, on this wall, burn them in the stone so that—if anyone comes after, if anyone is left to come, perhaps he will know—if he can read—of your suffering—and your foolishness.

If you had three grandchildren, what would you do?

41.

Every Flower That Exists First Opened In The Mind Of The Dreamer Of That Flower…

Maybe astrophysicists are zeroing in on the XLB236 of legend, ascribing to it a periodicity of 236 days, with pairs of solstices and equinoxes occurring 118 days apart, but the inhabitants of XLB236 view their efforts with a certain amusement because although they can see the gigantic sky-eyes combing the galaxy for cozy niches such as theirs, they have no fear of colonizers. They know that for Earthlings with such limited abilities the sea separating them is too different to cross.

XLB236's's summer solstice is the stuff of legends. On Flowerawakening eve, in the hedgerows, young ones of all five genders fill their sacks to bursting with ripening flower buds. At sunset they race over the greenswards to where their elders wait. Each elder extends a water cup, just large enough to hold one bud. Through the night each elder watches as the sleeping buds begin to stir, slowly unfolding their white, translucent petals one by one. In the deepest hour of the night, fully open, each flower reveals a trembling crown of stamens and at its heart, the inner tube gives off a scent so faint that, even though the elders barely sense it, it draws the night moth out of darkness. Rarely the moth stays long enough to plumb the flower's heart, but if it does, that elder knows: the time has come to walk into the hills, scatter the petals without looking, and lie down in that place.

Come spring, the young will find a new hedgerow growing

there. They will fill their sacks with the 100-year flower, from cactus nourished by their elders' bones.

42.
No See'ums II

Ninety-five days following the Fukushima Daiichi nuclear explosions and meltdowns, statistical evidence from the Centers of Disease Control shows that in eight cities in the northwestern United States (Boise, Seattle, Portland, Santa Cruz, Sacramento, San Francisco, San Jose, and Berkeley) infant mortality (i.e. deaths among infants under one year of age) has increased from an average of 9.25 per week to 12.50 per week, amounting to a statistically significant increase of 35%, whereas the total for continental U.S. rose about 2.3%. (Janette D. Sherman, M.D. and Joseph Mangano: "Is the Dramatic Increase in Baby Deaths in the U.S. a result of Fukushima Fallout?" *Counterpunch*, June 12, 2011).

But long before, in 1978, when Nikolai P. Dubinin, a leading Russian geneticist addressed the first International Congress of Genetics ever held in the Soviet Union, he had already observed that the percentage of children in industrialized countries born with congenital defects had *more than doubled* between 1956 and 1977 (Bertell, *No Immediate Danger*, 181).

In Tokyo today, each person inhales (or ingests) on average ten hot particles a day. In Boise, Seattle, Portland, Sacramento, San Francisco, Berkeley, and San Jose, five. Hot particles of Iodine 131 lodge in the thyroid; Strontium in the bones and teeth; and Cesium (134 and 137) in soft tissue, including the heart (Karl Grossman. "Fukushima and the Nuclear Establishment").

Perhaps the human body can be seen as a small universe. Millions of micro-organisms lodge in the gut, which, like those living in the soils, break down nutrients, make life possible. Perhaps now, more people may begin to revise their notion that the human body—our gift—is encapsulated for all time against particulate infiltration. We will have no choice but to let strontium brighten our smiles and pulverize our bones, iodine guard against unsightly goiter, cesium fire our hearts in radioactive glow so each becomes her own religious shrine.

In each tiny body compartment we will carry a Fukushima snapshot like the family pictures we bury in our wallets: microscopic images of blasted landscapes, boats wedged on the roofs of houses, pipes twisted like pretzels, irradiated water pools floating in our cells, the robotic images of small holes eating into the planetary mind, eating into our own small planet's membranes, insuring that our own molecular death will be the supreme gift of mankind's advanced technology—and his greed—and no longer the result of shabby natural processes.

43.
Pencil Do-Si-Do

One hundred and six days following the nuclear catastrophe at Fukushima Daiichi, AP, the chief press agency of the United States, publishes the results of its year-long investigation of the 104 nuclear power plants operating in continental United States; a year of examining inspection reports, regulatory policy statements and government and industry studies along with test results, a year of interviewing managers, regulators, engineers, scientists, whistle blowers, activists and residents living near some 65 of 104 nuclear reactor sites, most of them located in the Midwest and East. In all, four decades of documentation were examined, four decades because originally these plants were designed to operate for forty years after which they were to be decommissioned (Jeff Donn. "AP Impact: Industry and NRC re-write Nuke History").

The AP report found when old, rusting parts were discovered, the pencils of regulators repeatedly went to work revising safety standards downwards. Reference temperature, a cozily obscure term marking the temperature point at which a reactor becomes so dangerously brittle it becomes vulnerable to failure, got revised upward 50 percent and then 78 percent above the original benchmark—never mind that a broken vessel—like Fukushima Daiichi's— could spill its radioactive contents into the environment.

With the dance of the pencils, softer tests were devised. Leaking valves went away. Poof! Cracked tubing went away. Poof! Failed cables went away. Poof! Busted seals. Poof! Broken nozzles. Poof! Clogged screens. Poof! Cracked concrete. Poof! Dented containers, corroded metals, rust, hard-to-reach underground pipes. Poof, poof, poof, like magic wands, the pencils made thousands of age-linked problems disappear.

Pencil dance steps are of two kinds: the sideways shuffle, where the pencils change their interpretations of the regulations, and the forward do-si-do where the pencils change the assumptions behind risk assessment. Shuffle, shuffle, do-si-do. And poof! It all goes away. The roads go away, the houses go away, the abandoned cars go away, the landscape goes away.

The 20-mile exclusion zone around Fukushima Daiichi is eerily empty. There are no people, no traffic, no sound, just the eerie shadow thrown by the setting sun of the lone videographer stepping over piles of earthquake and tsunami rubble, documenting this no-man's-land that can never be re-inhabited for 100,000 years. There is no way to decontaminate a zone larger than 17 times Manhattan Island. Only a pack of stray dogs—abandoned or lost by their hastily relocated owners—roams the empty roads. They, too, because of the density of their hair, can never be de-contaminated. They will have to be shot—like the dogs of Chernobyl.

44.
Tears, Unexpectedly

It is the summer solstice, one hundred and two days following our planet's nuclear catastrophe. I walk out into the sunlight of a perfect day. A taxi waits at the curb. It's ninety-seven days since my accident. I hear the music of Africa in my cab driver's words. "What country are you from?" I ask him. "Nigeria." He still misses it, he tells me although he's been here twenty-two years now.

My report is a good one: the swelling of my nose has nearly subsided. My hair, thinned by stress, has begun to re-grow. Physiotherapy will take care of the tension in my neck and back and will correct my sciatica. My doctor is visibly delighted with these results.

I know she's taken a role positioning her HMO to be the administrating organ for a national single payer health scheme—if ever the people of the United States are permitted such a luxury. Before leaving her office I ask her what she thinks this government ought to be doing to inform people about the fallout over the northwest coast? What advice does she think this HMO should give to its subscribers? Her face looks as though it is about to burst. And then, quite simply, she breaks into tears. Her mother has had a stroke. She is alone to care for her. She tells me she can't take on anything more just now. I wrap her in a tight embrace, conscious that I may be trying to heal my own physician. But can we heal the planet?

As I wait in the downstairs pharmacy for my name to show on the callboard I read Isabel Wilkerson's astonishing history of the black migration north. She describes how in my own XLB365, in Atlanta, it's the colored ladies—not the elders—who gather at twilight around their doyenne's porch. They sit sipping tea through the night, waiting for her night-blooming cereus to open. Around three a.m. they go outside. The flower is ready to display its crowned magnificence. The colored ladies sigh. They look for the best thing they know on XLB365, the image of the baby Jesus inside the corolla, but the doyenne herself denies ever having seen it (Isabel Wilkerson: *The Warmth of Other Suns,* 239).

45.
The Flooding That Isn't, the Plant That Wasn't

On the hundred and fourth day following the planetary nuclear disaster at Fukushima Daiichi the internet carries a number of articles covering the news blackout about the Ft. Calhoun Nuclear Plant's red alert. The media jumps to reassure us the plant was not in operation at the time of the Missouri River flooding because the fuel rods were being replaced. It conveniently sidesteps reporting that the radioactive waste, together with similar "spent" fuel rods from other area reactors, are stored in open pools there, pools whose back-up generators may be in the process of being flooded. A huge balloon snakes around the plant's periphery, bolstered by sandbags. The locals watch the water rise, and rise, and rise and stop just one foot below the balloon's crest. And then a trucker backs his semi into it and pop!

But El Presidente, doesn't want ugliness fouling the Pre-Election air. That the rising waters of the flooded Missouri have created a red nuclear alert at the Ft. Calhoun reactor won't advance his chances. Nor will the withdrawal announced yesterday of a pitiful 10,000 troops from a ten-year war score him many points. His country has been looted by his banker friends. His medical plan has been trimmed to non-existence by a dysfunctional Congress. And somewhere in the Midwest after robbing a bank of one dollar an old man waits patiently for police to arrest him. He can't afford medical treatment for a complex set of problems. But in jail, he's happy to have scheduled several medical appointments.

In the millions of foreclosed, boarded up houses all over America's suburbs, thieves have removed bathroom fixtures. They've ripped out copper plumbing to sell for scrap. With school-board members' complicity, municipalities are selling their school systems to corporate low bidders. The TSA has set up a facility in Texas to sell the gewgaws it confiscates from all the travelers streaming through the U.S. airport scanners to catch their flights. A (non-presidential) White House spokesman reminds a fawning media that public opinion doesn't matter any more. And, investors take note: the Port of Piraeus is up for sale as one of the conditions of the IMF bailout of Greece.

46.

If the Price Is Right: A Monoplay

One hundred and eight days after the Fukushima Daiichi meltdowns of G.E.'s Mark I reactors Wikileaks releases diplomatic cables revealing how U.S. Diplomats shill for U.S. Corporations.

It was just a tsunami, just an earthquake after all. A pretty big one all the same. From my radiant heat high rise outside Tokyo I scan my plasma screen, let it filter what it wants us to see of the ruins of Sendai. I turn it off. I don't want to see the twisted metal—all that's left of TEPCO's three exploded GE Mark I Reactors.

Maybe the American commercial officer, maybe he didn't know the design was defective. Certainly I didn't know—I was just involved on the business end—although GE's terms were pretty fishy—but technically I didn't know. Anyway, now I'm retired. Someone, someone down the chain will have to take the rap—not me.

A million yen? Not so bad if you think of it. Anyway, if you didn't take it, if your government said 'no,'—tariffs, trade sanctions, embargoes—a big headache. They make sure you understand. Doing business with Americans was no sinecure, let me tell you, with their ha, ha-ing and back slapping. You're never allowed to say what you think of their bad breath and their badly tailored suits. I earned my pension, every yen of it, the high rise, the Mercedes, the chauffeur. The bodyguards—not just for me, but for my wife and kids—whenever they're up for a visit. I'm retired now. Chances are pretty good the yen stops before it gets to me. But someone had to know it was defective. I was on the business end, I didn't really need to know, even if—much later on—my son made sure I would. Whistleblower-san, what's his name? Bridenbaugh! That's it. The GE Mark I. Called it "ten pounds of energy in a five pound sack." Just before he quit.

47.

Pandemonium: All You Need is 45 Minutes Warning

Most of the 104 continental U.S. nuclear plants were originally located in remote, sparsely populated areas, many of them close to marginalized populations. There were never any restrictions about people gradually crowding into those areas with their houses, families, children, cats and dogs. In the past 30 years, within even a minimal 10-mile evacuation zone, the population around these plants has grown 62% to some 65,000 people on average. Meantime, the older the rectors get, the more they show the predictable signs of aging. The area surrounding San Onofre, located in California's most seismically active zone (the Pacific Plate) has nearly tripled its population density to just over 3 million. It lies within a 50-mile radius of Los Angeles and San Diego. A catastrophic accident would impact California's agribusiness, and contaminate its—and the world's—food supply.

If a 50-mile evacuation order were issued for Indian Point, 34 miles north of New York City where two aging reactors are due to be retired in 2013 and 2015, close to 17.3 million people would have to be moved in the space of one hour. Except the evacuation guidelines only provide for the first ten-mile radius, and no more. The next ten miles, the "god-help-you zone," you evacuate on your own, anyway you can. The next 30 miles are the "education zone," the area where you get to tape over your windows and stay indoors, and figure it out for yourself.

Highway infrastructure has not kept up. On ordinary days narrow, winding roads and choke-point bridges cause rush hour traffic to crawl for miles—with no allowance for nasty weather conditions. While regulators have upgraded maximum power output for aging reactors 139 times, allowing units to run at ever higher reference temperatures, the size of evacuation zones stay the same. Following a 45-minute warning, no standards exist for how quickly people need to evacuate. Evacuation plans for all these populations are schematic at best. Emergency experts have made sure to relax safety standards. Anyway, how are regulators to suppose people will obey directions to stay put? How will plans work where emergency crews have never conducted a thorough evacuation simulation like they do in Japan? And why haven't they?

A 1982 report predicted 64,000 deaths at New York's Indian Point, in a study that so shook up the public, no other such analysis was attempted for decades. But the noose is tightening. In the struggle to prevent the re-licensing of Indian Point, Alex Matthiessen, president of Riverkeepers, an environmental group, stated, "If they applied any meaningful standard to evaluate the emergency plans of this nation's nuclear power plants, there would be no nuclear power plants in this country, at least not in populated areas." Another nuclear safety activist throws up her hands. "Picture me with my son on his BMX bike, and my daughter at dance class, multiply me by 100,000—and you have pandemonium."

But in less than half an hour—as happened recently—a Madison Avenue shopper managed to drop $55,000 buying 5 pairs of crocodile fetish shoes—shoes presumably not designed for hasty evacuations—one pair for each of her five houses, none of them presumably located anywhere near a nuclear power plant.

48.
Phillip Glass, Step Aside!

Replacing serial composition, the music world recently welcomed the birth of the sulfuric acid school. Derived from experiments where scientists took pianos that had been allowed to fall into a state of mild disrepair—although of course, the action was still intact—but perhaps not necessarily in perfect tune. They identified this grade of instrument as moderately functional.

Initial experiments required that the strings be evenly doused with sulfuric acid. A large recording array was wired to the sounding board. The sound produced by the disintegrating strings was carefully recorded. Vibrations were plotted on oscilloscopes.

Applications led to new experimental compositional techniques. A piano concerto composed exclusively of elements of acidic decay has been hailed as the first triumph of Acidism, the new compositional wave that made Serialism obsolete.

> *Andante:* The main theme of climate collapse is followed by absorption variations where the melting of the polar ice produces areas of darkness where the snow is no longer able to reflect the sun's heat. Methane gas begins to be released.
> *Accelerando:* Variation on the central theme where weight initially concentrated at the poles gets redistributed,

accelerating seismic activity of crescendoing magnitudes.

Allegro: Population explosion requiring increased need for food, clothing, shelter, and energy.

Presto: Industrialization variation, developing themes of globalization building into the

Cadenza: Feedback loop in a four-part canon: more population, more industrialization, more globalization, more need for energy

Climax: Nuclear energy theme. This movement is necessarily short, abrupt and brutish, ending in a resolution where themes of famine, flooding, fires, and megaquakes alternate, concluding with

Coda: Non gracioso.

Eight hundred thirty-five pianos were destroyed in the course of creating the concerto's seven movements!

49.

And Now, Ladies and Gentlemen:

Feast of Saint BombsBurstingInAir, July 4th and 116 days following the global catastrophe originating in Fukushima Daiichi with nuclear fallout now making its way throughout the world, firework displays light up U. S. skies like so many amusement park stand-ins for our DU-tipped cruise missiles and bunker busters, drone-targeted strikes, DIME bombs (dispensing phosphorous), and cluster bombs (otherwise known as anti-personnel devices).

Count Lafayette may not quite have anticipated the consequences of abetting what started out innocently enough as a Planter's Revolution. His own national anthem, "The Marseillaise," wouldn't appear until thirteen years later:

> Come, all ye children of the nation, the glorious day is upon us. Tyranny's bloody banner is raised against us. Hear the sound in the fields/the battle cries of armies/sent to slash the throats of your brothers and your sons. To arms, Citizens, form your battalions. March on, march on, that

tyranny's impure blood may spill in our furrows. Liberty, dear, dear Liberty, join our struggle so that in death, your enemies may see your triumph and our victory.

A terrorist anthem if ever there was but clumsy all the same.

Of local interest: On the 115th day following the Fukushima Daiichi planetary disaster, an explosion and fire rocked block 1 of the Tricastin nuclear reactor located just north of Marseilles.

50.
A Game of Marbles

My late friend, Padraigin, used to talk about the London blitz. As a child of three, the blitz meant nothing to her. The bombers were a half-hour away, but she paid no attention to the sirens. She was on her hands and knees, trying to stop all her wayward marbles from tumbling down the gutter. Suddenly, she felt two hands grab her. A strange man rushed her into an air raid shelter and probably saved her life.

My Bay Area cousin plans to drive from outer San Francisco across the Bay Bridge and into Oakland. She's promised to deliver some blueberries for an outdoor barbecue she's too busy to attend, but she plans to drive 30 miles to deliver those blueberries, releasing 14 pounds of carbon dioxide along the way. She loves her daughter, she worries all the time about her future, but she has to drive those 30 miles.

My San Diego cousins have 11 grandchildren. They love their grandchildren, they have time for nothing else, but they are planning a vacation drive to Vancouver and then into British Columbia, camping all the way. They love their children and their eleven grandchildren, but they plan to drive 7,000 miles, releasing approximately one ton of carbon dioxide each way.

I reflect what these small pieces of the human puzzle tell me. We may not be suited to this planet, our mind not attuned enough to understand where we live. We are still busy collecting the small glass globes in the gutter because the large globe on which we live is way beyond the scale where we can see it, where we can stop it rolling down the drain.

51.
July 13. Shadow Play

A man is talking, talking to an interlocutor, a reporter maybe, shown only in silhouette—TEPCO has barred its rank-and-file employees from speaking publicly. If his identity were known, he might lose his job. He's on a hazard crew assigned to clean up the remains of what had once been the Fukushima Daiichi reactors. Why is he here? Why is he doing a job that will probably cost him his life? Is he aware that once inside the body, radioactive particles bombard neighboring cells until cancer develops? He says there are 1,000 workers cleaning up the site. They sleep in dormitories to be near the plant. They wear respirators, hazmat suits. In the heat of July, some workers succumb to heat stroke. They can't wipe the perspiration that accumulates under their respirator masks because they cannot touch them while they work.

He began his work life here. He took part in building this installation at the edge of the sea, in a seismically active zone, although at the time it was believed not to be seismically active. For him it is his flagship. He will go down with it. It's not as if he's sacrificing a young life. He's in his mid sixties now, no longer young. It is the debt he must repay, no more, no less. He is responsible. He is responsible for the great harm he has caused, harm to the soils, and the land and the villages dotting this coast. And to the people who can never return here, to the fields they have lost and the houses. And the silent streets, and the heaped up detritus; for the refuse that lines the highway for miles and miles, old sofas piled with stuffed animals, along a stretch that used to be dotted with amusement parks, now cluttered with the equivalent to 23 years of leavings, some of it carefully sorted out into recycling piles: metal and all things metal to be resold for scrap, if it is not radioactive; broken up concrete, once the rebar has been separated out, to be sold for rip rap—if it is not radioactive; piles of wood from house siding, from furniture, from uprooted trees that will go up the chimneys of the incinerators—three of them—being built now for just this purpose.

He will stay here and do his duty until he falls or till it's done, whichever comes first. Because he is responsible to the people who lived here once, to the prefecture, to his country that has suffered this terrible blow—because of him, because he helped build this reactor, he was working here when it went on line. Because it brought about the great disaster of his time.

The man will think of his wife perhaps, as he turns over in his dormitory bunk. He may think of his children, wonder where they are sheltering. In some stadium perhaps, or school gymnasium, sleeping on the borrowed mats furnished by the state—or some benevolent organization. He may think of them lining up day-by-day to receive food rations on trays supplied by a neighboring prefecture—outside the exclusion zone. He may think of his grandchildren and wonder briefly what will become of them. Will they carry this uprooting with them throughout their lives? How will its mark stamp itself on their spirits? What of it will they hand down to their children and their children's children? Will its eddies still be felt unto the seventh generation? Is that what time looks like, fissures in bone and sinew of generation on generation? Will he lie awake in his bunk pondering these things, listening to the night breathing of the dreamers who share his station, in the building where they sleep, where they will rise again before the heat of day, at 6 am, or maybe even 4, to suit up again, in suits he knows don't stop gamma rays, once more to go into the wreck, once more try to make sense of the twisted pipes, the conduits, the torn up cement. Where they will sweat under their respirators in the unbearable heat and humidity of summer, where all thought of snow and ice and the freezing temperature of winter can't yet be imagined.

Time passes. Like a slab of liver on a butcher's block to be sectioned, first one lobe, then another, into days, weeks, a fortnight, months. Sectioned in smaller and smaller pieces until it disappears.

52.
Forget the Planetary Mind

Taken from the documentary "Reimei," the front page of *The Japan Times* on July 15, day 126 following the disaster at Fukushima Daiichi features a set of excavation stills from the early 70s when the site for Fukushima Daiichi was being leveled, its red earth tamped down, flattened by the great machines, a square excavation, of no more interest than a cake pan. According to Katsumi Nagamura who worked for TEPCO at the time, Fukushima was built on what was once a hill, a hill 35 meters (some 108 feet) above sea level, which TEPCO reduced by 25 meters to 10 meters (30 feet), confident a tsunami would never hit.

GE had never before designed a reactor to be located at the ocean's edge. But following GE's directive to the letter, TEPCO excavated a basement 14 meters (42 feet) below sea level in which to install diesel-powered generators, the back-up system that would insure no interruption to the cooling system pumps, pumps that in an emergency would prevent fuel rods from overheating and melting down. At the time regulators and seismologists were sure an earthquake of 9 Richter magnitude and the massive resulting tsunami would never hit the plant—let alone the planet. Why would it? A plant that represented the summit of human technological advancement, the ultimate Temple to Science in an area, which wasn't even seismically active! In the 1960s GE—and TEPCO—had no way of foreseeing climate collapse. They couldn't imagine that 50 more years of carbon emissions would melt the polar ice, or how the shifting loads would result in megaquakes. They left the planetary mind out of their designs.

53.
How to Destroy the World One Hindsight At A Time

Ah, ah, ah! 129 days following the planetary disaster of March 11th Yukiteru Naka shakes his head. Naka is a former GE employee who took part in designing and later operating Fukushima nuclear reactors units 1, 2 and 6. "As early as 1980 I started having my doubts," he laments—now. "Why were the back-up generators located in the basement, and so near the DC batteries?"

Ah, ah, ah, ah! These are the tears of all the weeping middle ranking engineers who say—now—they had the same concerns.

Boo-hoo-aha-aha! First in charge Tadaharu Ichiki, former nuclear plant designer at GE-Toshiba in the late 1960s is so sorry—now—that he followed GE's design to the letter, but the TEPCO Shogunate left all design responsibility to GE. *Put the turbines in the basement, never mind if it's only designed up to Class B specs for earthquake resistance and that it's not water-tight. There's never been an earthquake here.* Never mind that in neighboring villages, long lost stone tablets, many of them hundreds of years old, marked the high points of centuries of tsunamis.

Uhn, uhn, uhn. This is the demurral of Masatoshi Toyota, TEPCO's former senior vice president, who with other executives oversaw Fukushima's construction. He *never* knew—until March 11th—that the diesel generators were located in the turbine buildings! He says—now—he would definitely have changed all that. (Except that under national regulations back-up generators had to be designated Class A for quake resistance. Design usually housed them in tightly sealed super-robust reactor buildings.)

And supposedly none of them knew—following a huge U.S. campaign to sell the Japanese on nukes— that when GE sold the Mark I boiling water reactors to Japan, already then GE knew they were defective. Three design engineers had quit in disgust; and one of them referred to the Mark I as "ten pounds of energy in a five-pound sack."

54.
Getting It Wrong

"Civilization's" gift seems to be Getting It Wrong starting from the time humankind passed from nomadic pasturing to agriculture as a basis of survival, and animism was replaced by The World's Great Religions. Agriculture implies land tenure. Aggregates of land—and water—come to favor the few at the expense of the many. The dawn of agriculture eventually led to the establishment of the first city, Çatal Huyuk in Turkey around 6,500 B.C. Soon it became clear that the city model could not support life without importing things from the exterior. New aggregates of distribution entered into play. We have seen this concentration of power and resource reach its apogee in our century, where the elites of one nation exert dominion over the entire earth in the service of amassing yet more aggregates of wealth.

This is not a board game of monopoly. The frenzy with mining the earth any which way to feed a devouring appetite for energy is leading to the poisoning of the planet's water, its seas, its soils and its air. And now with the deadly cocktail of nuclear plants—over 400 of them on earth—mixed with megaquakes brought on by climate collapse, survival of all life on the planet hangs in the balance.

Driving across continental United States may not be the best way to cool the planet or even the best way to connect the dots of climate collapse, but just such a driver witnessed in large what is not given to many of us to see first hand. Janet Redman writing for Other Words on day 129 following the planetary disaster at Fukushima Daiichi describes such a journey. Above-record scorching heat through the Eastern states, heat intense enough to melt roadways, flooded cornfields in Iowa, rivers overflowing their banks farther west into Nebraska, the detritus of flooding and hail storms freakish enough to damage fleets of airplanes, and in the Rockies, fires, some of them reaching into the boundaries of Los Alamos nuclear waste repository, and into the surrounding canyons, just about guaranteeing that, with winter rains, isotope-contaminated soils will wash into the watershed to contaminate the water table (Janet Redman. "Connecting the Climate Dots Across the Map").

55.
Today's Headlines

Two of today's headlines one hundred thirty-two days following the global industrial catastrophe originating at Fukushima Daiichi:

The U.S. Department of the Interior Considers Making a National Park to Memorialize the A-bomb: several in Congress support the idea

and

One in Sixty-Six Americans is Psychotic

Plans are on the table for a national park dedicated to celebrating the Manhattan Project. The race will soon be on for an award-winning design—parking lots paved with a light shade of concrete, light enough perhaps so that images of the victims' photoshadows can be burned onto each space. Barbecue pits designed in the shapes of those carbonized—somewhat on the order of Rodin's Gates of Hell—but any resemblance to any charred persons living or dead will be purely coincidental. In the artificial lakes and ponds, pneumatic recreational flotation devices inflated to resemble people who piled into the rivers to quell the burning of their flesh. Landscaping trees and shrubs trucked in from dying forests to be attractively relocated. Wildflowers seeded from all the lovely species that bloomed suddenly over Hiroshima in the weeks following the blast. Dedication ceremonies will be scheduled for August 6, 2020 to mark the seventy-fifth anniversary of America's Great Achievement and the inauguration of America's Great Achievement's Amusement Park. You can tour other monuments glorifying nuclear annihilation at Hanford Reservation, Rocky Flats, Colorado, the National Wildlife Subterfuge at Fargo, N. Dakota, which includes the ICBM "Ronald Reagan" launch center, or the Minuteman Missile National Historic Site in South Dakota. To round out the tour, other "parks" are planned for Los Alamos, Oak Ridge, Tennessee, and even more in Hanford.

56.
And In The Fluid World . . .

"What brings you to take up linguistics?" my young classmate asks me. I see a young man with the liquid brown eyes and hair and beard of a Knossos fresco. He must see an old lady under a sunhat fit for a junkman's mule.

"Because I'm asking the question: why is the world construed as it is?"

"How so?"

"Progress. Technology. Full speed ahead. And where have they led us? Our highest achievements, but without the mindset to understand that—highly sophisticated or simpleminded—we all live here. On a shared planet. This is our home."

"So you're studying linguistics...?"

"Yes, because the world doesn't have to be construed the way it is. Other beings may see things differently. Native American languages—some of them—reflect that. And natural history. Dolphins and pseudo-orcas may have things to teach us."

"How?"

"Ah, that's a longer conversation. It has to do with what you might prefer: learning how to fetch and tell apart yellow, red and blue balls or yellow, red and blue washcloths, or learning how to trust."

"Why?"

"Because trust is what cetaceans could teach us."

Dolphins have been found to share a sense of sexual play with bonobos. In his writings, John Lilly, polymath explorer of interspecies communication, records an "experiment" conducted by one of his "researchers" sometime in the 60s. Between "training" times with "Peter", Lilly's "researcher" stroked her experimental subject's penis to the point of bringing him to shuddering satisfaction.

She taught him sufficient recognition of basic English that he was able to distinguish between, and fetch, yellow, red, and blue washcloths and balls. If he showed signs of confusion at first, she delivered a smart rap to his snout until it became ulcerated.

"Peter" showed eagerness to "learn," but not always along his "experimenter's" terms. During the few months "experimenter" and dolphin shared the same tank, repeatedly "Peter" clamped his jaws over his "experimenter's" calf and gently ran his razor sharp teeth along its length. Dolphin jaws are known to be strong enough to snap a shark in two. This behavior, far from being encouraged, was rewarded with more raps to the snout.

One day, "Peter" approached his "experimenter," a colored ball propping open his jaws. Slowly, deliberately, he ran his teeth up and down her calf, clearly communicating that his intention bore her no harm. Perhaps he was trying to reciprocate her rather blatant courtship gestures in a more gentlemanly way.

In the wild, dolphins swim in pods, usually led by a matriarch. They fish collectively, and they observe rituals of sharing. The matriarch might observe a school of mackerel swimming at some depth below. At her signal, a "pilot" will lead the pod to surround the school. They will swim in ever tightening circles around the mackerel until they have massed them in a dense ball of sushi. Then, one by one, they take turns, each eating their fill, beginning with the youngest individuals. They will do this until all members of the pod have eaten to their satisfaction. Captive dolphins have been known to refuse some "trainings," perhaps because they consider them demeaning (Jim Nollman: *The Charged Border*, 34).

57.

Why Take A Space Ship To Get Somewhere Else, When A Rattletrap Bus Gets You To Heaven? And It Only Costs A Dollar

On July 22, after class lets out, I take the bus up the canyon to my own personal XLB365 (37.77'18" N X 122°16'22" W) the dollar fare mundane as a locker key, the Berkeley University Botanical Garden (where the jack in the pulpit grows three feet tall), paradise. This one hundred thirty-third day following the explosions at Fukushima Daiichi is glorious, reminder of days when summer skies were cloudless, the air gentle, still but for the waving of ferns and the wing sweeps of butterflies drinking in the noonday sun. I find what that day becomes my favorite bench to sit and read and ponder. I look up from my book from time to time, catch the sharply etched boundaries of leaves solarized against the sky, the desultory flight of a yellow swallowtail, the hurried flutter of a fritillary. Let the light visit me disguised as pine needles, translucent ferns, let the air caress the shining grasses, shake the aspen leaves. Breathing. This moment. Still. Like no other. Come to its perfect stop. This planet, this and no other. This one—and no other—for its breathtaking perfection, until there is no bench, no book, no woman sitting there. All, all is as one. *There is no place more wonderful than this. There is no place more marvelous than here.*

Back at the bus stop, my bench mate laments we may just have missed the bus. As we wait, she shares with me how she feels living on our endangered planet. She's shed her distractions: cell phone, TV, gone cold turkey from watching half the day and through the night. Now she's content listening to her jazz in the fixer-upper house she shares with her mother, and a grandmother who still knows everybody no matter where she goes.

Above us, the crowns of tall eucalyptus sway slightly in the wind. I shield my eyes from the blinding light of the afternoon. Beyond the tree line a lone hawk takes to the air, spiraling lazily. What does it feel like to be up there, climbing ever higher, riding the air currents like a spiral staircase? Soon it reappears with its mate, slowly circling, spiraling ever higher, tightening their circle, and suddenly vanishing.

58.
Uninhabitable Zone

His picture is captioned on the front page, the image of a man, face bowed, features dissolved in tears. One hundred forty-two days following the catastrophe on planet earth, Professor Toshiso Kosako resigns. He's an expert on radiation safety, and the second minister appointed since March 11 to oversee coordination of Fukushima cleanup efforts. He accuses Prime Minister Kan and Japan's Nuclear and Industrial Safety Commission of violating the law regarding radiation hazards and permitting children living near the destroyed reactor complex to be exposed to doses of radiation equal to the international standard for adult reactor workers. "There is no point for me to be here," he tells the press. A government minister weeping unabashedly, announcing that for the children of Fukushima to be exposed to radiation above threshold levels is unconscionable. He cannot save the children from government fiat; he cannot announce publicly that one-eighth of Honshu has become uninhabitable—in an island nation whose population density is already one of the highest in the world. He cannot say that already 500 children show signs of irradiated thyroids.

What would it mean if one-eighth of the United States became uninhabitable? Say roughly the territory from the Appalachians to the Eastern Seaboard where some 53 reactors, 14 of them in areas of high seismic activity, and 8 with records of "near misses" are located, and where 30% or roughly 112,640,000 million of the U.S. population lives. Would our president bow his head? Would he weep unabashedly? Would he apologize? What would the 90 million people sleeping on cots, under army blankets packed into school gymnasiums and armories and churches think seeing the announcement of their president weeping on the television screen? (For an interactive U.S. reactor map go to the Union of Concerned Scientists' http://www.ucsusa.org/nuclear_power/reactor-map/embedded-flash-map.html.)

59.
The Scale is Relative

I cup in my hand the badly washed bowl that held yesterday's marinade one hundred forty-nine days following our planetary catastrophe at Fukushima, during the week the world marks the anniversaries of Hiroshima and Nagasaki, one of many days humankind is hard at work destroying the natural world. Something still clings to the surface, something that has no business being there. I scrape it loose. What can it be, this stubborn piece of detritus? I hold it up to the light, a dull, translucent flake, bean shaped, fragile as paper, no bigger than my fingernail. What can it be? with ridges in its midsection, like the shadowy imprint of a partly opened fan?

A fish scale! its give and its resistance designed to shield and speed its wearer's passage through the water.

Can the great world's perfection shine in a derelict fish scale, in the iridescence of a peacock feather, the wing of a butterfly, in the smallest grain of sand?

"It's our garden—this world," I tell my friend.

"You see it that way just because you happen to live here," says my friend.

60.
Sixteen Women and Four Men

One hundred fifty-one days following our planetary nuclear catastrophe, on August 9, sixteen women and four men are arrested at the west gate of Lawrence Livermore National Laboratory, where 89% of its more than 1.2 billion dollars will be spent on nuclear weapons in the coming year. I am one of them. Why would I go, I ask my friend and colleague. This is an annual ritual. It does nothing to rid the world of nuclear horror.

You need to go, he counsels, not because it will rid the world of nuclear devastation, but because if you don't it will rid the world of your conscience. I go.

We stand in the early morning sunshine at the gate, some one hundred of us. We listen to Saori Matsuoka, a young Japanese exchange student, read a letter from her aunt in Japan: "our meat, our vegetables, out fish, our rice are contaminated. We have no food to eat that is not contaminated." Her aunt's daughter asks: "Mom, it's true, isn't it? We have no hope now?"

Lined up, legs akimbo guarding the fence, a row of Lawrence Livermore National Laboratory security personnel model the latest camo, boots, provocative black garters on their right thighs holding their gun holsters. They will not reveal to whom they report, but some of them have not removed the Velcro-bound insignia from their sleeves. Alongside them, personnel from the Alameda County Sheriff's office in plexiglas helmets, padded in navy blue body armor worthy of a samurai, take turns practicing reading the order to disperse. I keep my eyes fastened on their boots. I plug my ears as an air raid warning goes off. It's the signal for us to hit the deck. We lie immobile, a collective die-in on the asphalt. Those not risking arrest draw chalk outlines around our bodies. We do not disperse. One by one we are helped from the pavement, conducted through the fence, our pockets emptied, our few possessions placed in a plastic bag. Then we are handcuffed. All this occurs in an atmosphere of exemplary courtesy. They are solicitous of our old bones, our frozen shoulders. They oblige by handcuffing some of us with hands in front of our bodies in deference to our age—one of us is 85, with 27 arrests to her credit. (She keeps track at her grandchildren's request.) One by one we are very ceremoniously led to the van. When the van is full, it rattles down the road, taking us to the holding pen. We are playing at rogues gallery. Each by turn is escorted out in order to be photographed holding an improvised 8 1/2 X 11 sign with our names and dates of birth. Under the corrugated roof, we are processed one by one: name, address, phone, social security numbers. And fingerprinted. We are asked to sign our intake documents. I take inventory of the storage shelves: cartons, spools of electric wire, enough stuff here to cover an elementary school's budget for one year. We are returned to the holding pens. An hour later, all of the 20 have passed through. It's a slow process this year because budget cuts have reduced all personnel, but still there are many more of them than there are of us. We are all affability; this one tells me he's been in the military all his life, that one—Rodriguez—is Latino like me; Garcia is half Mexican, half Apache, but doesn't speak his mother tongue. We trade the time of day, despite our contrasting wardrobes, brothers all the same. This ritual takes all morning. Once the last of us is released, we walk the quarter mile back to our parked cars.

In Japan radiation levels are reported higher today than ever before. Evacuation plans in place since the 80s are revealed not to have worked. None of the March 11 weather readings ever reached Prime Minister Kan's desk, and he was the only official empowered to give the evacuation order. Local Mayors were all left in the dark. People of Fukushima were ordered north to Tsushima. They walked or drove directly into the radioactive plume. Far from being evacuated, children sheltered in Karino elementary school ten miles from the explosion. They ate rice balls served in the open air as the radioactive plume passed directly overhead. Other townspeople were told to shelter there for a period of some ten days before being evacuated at last from this highly radioactive area.

At the LLNL gate, the young Japanese student reads the letter from her aunt: "Many people who were exposed are shunned now—in the stores, in the street. They cannot shop in the food markets or eat in a restaurant because people fear they are contaminated. They are discriminated against at every turn. People don't want their sons or daughters to marry anyone from Fukushima because they are afraid their grandchildren will be born with birth defects resulting from radiation exposure."

In the U.S., the administration has just approved a multi-billion dollar loan guarantee for Reliant Energy (one of El Presidente's campaign's heaviest bundled contributors), to construct a new nuclear plant. Reliant Energy will partner with Tokyo Electric Power Company, TEPCO—the very same company responsible for building the defective reactors at Fukushima Daiichi—to build it. Further plans are on the drawing board to manufacture 100 drones specifically designed to carry nuclear weapons.

Twenty people are arrested today, sixteen women and four men, because they don't want their consciences to disappear.

61.
Flickerings of Fear

Where were you when Chernobyl blew? Do you remember? Where were you when the earthquake hit Fukushima Daiichi and caused it to blow? Where were you?

From the time of Chernobyl, when you fell to the ground in horror, and clasped your partner's knees, you knew this would come. You knew what it meant. You remember wave upon wave of fear, of terror leaping above the water line. Perhaps now you imagine you are sailing a becalmed sea, no winds, no waves, as calm as the Pacific Gyre with its plastic soup the size of Texas. Then you remember—for an instant—and for an instant fear flickers again just above the surface calm. And you dismiss it. You're good at pushing it back: under the bed, inside the darkened closet. You tell yourself San Onofre sits on an earthquake fault six hundred miles away. If there's a disaster, you won't be affected. Diablo Canyon sits on a fault three hundred miles away. It will not affect you. It will only affect…and you play god-roulette. The roulette of Abraham arguing for the lives of the Sodom-Gomorrans. How about if you spare 5,000,000? Is that a 'yes' I hear? How about 4,000,000: still a 'yes.' Now, how about 3,000,000? Will you spare Sodom-Gomorrah with a 3? A 2? And now you're down to one, and what happens when you discover that that one is your son?

God and Abraham are old stories now. What reigns now is *sauve qui peut,* dog munch dog, each man for himself, and damn the women and children. You're in this for the long haul. But you know that 50 miles away, at Lawrence Livermore, located over an earthquake fault, 3,080 pounds of plutonium are stockpiled and nuclear waste is stored where it contaminates the soils, the water and the air. And you know there's a small reactor at Davis, and behind Haas gymnasium, just seven blocks away from where you live, all the nuclear waste from Lawrence Berkeley Lab is stored. And you laugh at yourself for imagining that you could escape.

Escape where? You live in the U.S. of A. Do you imagine early warnings and evacuation plans would work any more efficiently—or more transparently—than they did in Japan? Or in the USSR at the time of Chernobyl? Do you think your government behaved any differently in 2005 at the time of Katrina when the entire black population of New Orleans was left to drown?

Much as you might like to think it, you are not immune. You would lose your dwelling—and your composure. You call it dwelling because it has all that good niceness about it: nice tablecloths, nice candlesticks, nice art on the walls. Others—those less fortunate—might lose a house, subsistence dwellers might lose a roof. But all those unwashed, uncombed homeless men—too many of them veterans—walking with their bedrolls and their shopping carts, now they would become king! Because they know how to do this one better than you ever could. They've served their long and miserable apprenticeship. You'd exchange your cuisine for their scraps. Everything you touched would be radioactive, what? would be? It already is! You were caught in the rain March 25th, fourteen days after Fukushima Daiichi blew, your coat got soaked. Your shoes got soaked. You brought small traces inside, inside your home. Traces contaminate your rugs, your sheets and towels, your clothing. You drink water from your tap. That water comes from a reservoir where, borne on the trades, the Fukushima Daiichi plume passes overhead every day. Do you imagine your water is not radioactive? And the vegetables you eat, and the fruit? Do you imagine they are grown in hot houses?

Despite your government's sealed lips, have you forgotten the hard Center for Disease Control statistics? In eight northwestern cities infant mortality has spiked 35% since the Fukushima Daiichi disaster. Your city is one of them.

62.

Nuclear War in Black Face

We sit in cushioned church pews watching "The War You Never See," a film by journalist John Pilger, which American censors prefer us not to see. We cringe at images of wounded Iraqi and Afghani children, their small bodies buried in the rubble, victims of the smartest, most efficient bombs the world has ever seen. We register the statistics of mounting civilian casualties: in WWI, 10%; in WWII 20%; in Vietnam, 50%; and now in the pan-global U.S.-sponsored Wars of Empire, an efficient 90% of all casualties are civilian. We watch our fashion plate "president"—our own dusky Caligula—strutting the campaign runway as the flash bulbs pop, his million-dollar fashion plate wife on his arm—who has deployed even more troops, fired even more missiles—missiles tipped with depleted uranium—deported even more immigrants, sanctioned even more targeted assassinations, than the boy tyrant who preceded him.

We watch Pilger interview Cynthia McKinney. She speaks slowly, deliberately. Her pauses yawn almost as if she wants to sleep. "Black people by and large don't want these wars…It's as if the wars were being conducted…in Black Face."

All these one hundred fifty-five days following our planetary catastrophe of March 11, I have pondered how the race, soon to number 7 billion, has backed itself into this End Game. I ask myself: could this world, could our species, housed in these awkward bodies, walking upright, thumbs in appositive relation to our fingers, could this species have construed our world any differently? Was another way possible?

Said Chief Seattle to President Pierce: "Every part of this earth is sacred to my people. Every hillside, every valley, every clearing and wood, is holy in the memory and experience of my people. Even those unspeaking stones along the shore are loud with events and memories in the life of my people. The ground beneath your feet responds more lovingly to our steps than yours, because it is the ashes of our grandfathers. Our bare feet know the kindred touch. The earth is rich with the lives of our kin."

63.

Fitting It Together

What if you had to map the landscape on your body? You would have no words to say the river ran to your left. It would have to run to the west, or the south or east, or north, depending on which way you faced. You might talk about your heart racing toward the east, or your ear aching to the north. If you did, would your sense of where you live be more connected to the landscape?

Supposing you couldn't spotlight an individual of any category. Supposing you said deer and sheep and grizzly bear, to mean the generic sum of all deer, or sheep, or grizzly bears. Supposing in order to pinpoint an individual you had to add a small suffix (a 'that') at the end of the generic word you used (bear-'that'). Would you be inclined to see the world whole in all its parts?

Supposing you didn't have the word 'he,' a third person singular generic word that marked for gender. Suppose you had a gender-neutral word that could mean man or woman. Would you be inclined to see the women—and the men—in your world differently?

Or supposing you could never claim possession of another living being. Not even a baby. You would have to say the baby lives with me. You could never refer to your wife, or your husband. You would have to say: I am wived; or I am husbanded. Would you be inclined to endow all living beings with a particular sense of respect, and accord animals and people you take for granted each their separate dignity?

If your son or your daughter took sick, supposing you had to say: I am sick with respect to my daughter, or my son, would you be inclined to feel more keenly someone else's pain?

Supposing any event you described to anyone would have to carry information about whether you actually saw it with your own eyes, or whether you inferred it, but didn't directly see it, or whether it was only told to you. Do you think you would be more careful about any claims you made?

A language with such attributes still exists, although a 1990 U.S. census listed only 6 remaining speakers. Their numbers originally were decimated by malaria, and massacres—notably by Fremont and Kit Carson—and poisonings by White settlers eager to invite them to a Friendship Feast. But before contact (A.D. 1770) Wintu speakers may have numbered anywhere from three to five thousand people inhabiting the Sacramento, McCloud and Trinity River watersheds (see Leanne Hinton: *Flutes of Fire*).

In the distance I make out a familiar figure walking toward me, his white cane testing the ground. He's a retired professor of linguistics fluent in at least one of the languages spoken by indigenous people in Northern California. I've met him some time ago, but ever since, whenever I give him my name, he always knows exactly who I am. I run my Wintu hypothesis by him. "If you had to specify clearly between direct observation and hearsay, wouldn't you be less inclined to lie?" "You'd be an even better liar," he responds, roaring with laughter.

64.
Lying Behind a Convenient Tsunami

Headlines of August 17 in *The Independent* (UK) reveal, "Explosive Truth Behind Fukushima Meltdown." The sequence of events is important, because *it turns out containment failure, and not the tsunami,* was the original cause of the disaster. But by stressing tsunami damage, TEPCO—and General Electric—have diverted attention from the main problem: the GE Mark I BWRs are defective. GE knew they were defective when they sold them to Japan (David McNeill and Jake Adelstein. "The Explosive Truth Behind Fukushima's Meltdown," *The Independent,* UK).

After the planetary catastrophe at Fukushima, why did TEPCO consider reducing its skeleton workforce to 50 persons and walking away? Two quakes followed the 9 Richter quake. The 9.0 earthquake of March 11 was responsible for destroying off-site power at Fukushima. Of that, there is no doubt. An NRC FOIA release contains text of a news release dated March 13 from the Japanese Ministry of Economy, Trade and Industry (the nuclear energy oversight body), which states:

> A radiation level exceeding 500 microSv/h was monitored at the site boundary (15:29, March 12). A large motion occurred due to an earthquake with close epicenter and a large sound was issued near Unit 1 and smoke was observed.

Why does the media remain silent about the 6.1 aftershock that occurred March 11 with an epicenter of 10 miles away, and the 6.4

quake on March 12 at 3:29 pm, with an epicenter of less than 30 miles away, exactly 7 minutes before Unit 1 exploded? *Hawaii News Daily* picked up the story on Day 226, October 23:

The Tohoku Tsunami Spin Take Two: The Missing Quake that caused the Explosion of Unit 1

> If there is one thing the whole world [thinks it] knows, it's the claim that the tsunami was the last straw in destroying the backup power that resulted in the reactor meltdowns and spent fuel pond crisis. There's a reason why every nuclear expert on the planet wanted the world to hear this and that's because there is also the claim that Japan is the only place such a massive tsunami could take place! Actually there is more than a bit of debate about this, but there was at least one other event they have completely censored from coverage anywhere: **the earthquake on March 12th** (Abalone Alliance Clearing House. "Fukushima FOIA: The Aftershock (Quake) That Blew Up Unit 1").

The media remains silent about earthquake damage, blaming the disaster on the tsunami, because it serves to protect the interests of the nuclear industry: it doesn't want you to know that the cause of the disaster was structural reactor failure. By blaming the tsunami, it obscures the truth.

Until now, TEPCO and the Japanese government have not established much of a reputation for truth-telling. "It was an unforeseen disaster," TEPCO's then-president said later. "There has been no meltdown," Yukio Edano, a government spokesman, stated shortly after March 11. Now, one hundred fifty-nine days following the catastrophe, we know that meltdowns were happening even as Edano spoke.

Prime Minister Kan's celebrated helicopter trip to Fukushima in the early morning of March 12 is the ultimate *Rashomon* in a web of TEPCO-related *Rashomons*. Why did the PM fly to Fukushima? Was it to force the plant manager, Masao Yoshida, to open the vent on Unit 1 to relieve the pressure build-up? Or did his trip amount to a government inspection, which delayed any effective action from the time he arrived at 7:12 a.m. and 9:04 a.m. when his helicopter took off again? (Remember, March 12 is the day Unit 1 exploded.)

According to the *Yomiuri Shimbun*, Kan's junket caused a delay, interrupting TEPCO's emergency measures and causing Unit 1 to explode.

According to Evan Osnos, writing in the *New Yorker*, Kan's helicopter intervention was the stuff of a Hollywood thriller. He portrays Kan as a hero-to-the-rescue, bullying TEPCO's management to vent built-up hydrogen gases directly into the atmosphere. According to Osnos, only the Buster Keaton antics of the plant manager and staff caused fatal delays that resulted in the explosion of Unit 1.

But the *New York Times* offers a more nuanced account. As Minister of Health in the 90s, Kan uncovered collusion between government bureaucrats and pharmaceutical company officials. The two parties had been keeping secret the fact that his ministry had been using blood tainted with HIV. Because Kan distrusts bureaucrats, "mutually suspicious relations between the prime minister's aides, government bureaucrats and [TEPCO's] company officials obstructed smooth decision-making (Martin Fackler. "In Nuclear Crisis, Crippling Mistrust"). *But all this speculation functions as distraction, diverting the public's attention from the main point: The GE Mark I boiling reactor was designed to fail.*

Mitsuhiko Tanaka, a former nuclear plant designer, identifies the accident of March 11 as a loss-of-coolant accident. "The data that TEPCO has made public shows a huge loss of coolant within the first few hours of the earthquake. It can't be accounted for by the loss of electrical power. There was already so much damage to the cooling system that meltdown was inevitable long before the tsunami...." He reviews TEPCO's released data. "At 2:52, just after the quake, the emergency circulation equipment of both A and B systems started up, something that could only happen when there's a loss of coolant. Between 3:04 and 3:11 the water sprayer inside the containment vessel was turned on. Such an emergency measure is only taken when other cooling systems have failed. All these indications point to a condition of meltdown before 3:37 when the tsunami knocked out the electrical systems."

Nine days before the meltdown, the Japanese Nuclear Industrial Safety Agency warned that TEPCO had failed to inspect recirculation pumps and other critical equipment. It ordered TEPCO to inspect and perform repairs, but as of June 2, there is no record of TEPCO ever having filed its report. Problems with fractured, deteriorating pipes and the cooling system in general had been recognized for years. Now we are beginning to hear from the people who actually worked on site at Fukushima, maintenance people and engineers. Several of them have requested anonymity because they still work there. They report that serious damage to piping and to at least one reactor occurred before the tsunami ever hit. A technician in his late 30s, who was on site at the time of the first earthquake, recalls: "It felt like the earthquake hit in two waves. The first impact was so intense you could see the building shaking, the pipes buckling, and within minutes, I saw the pipes bursting. Some fell off the wall.

"Someone yelled that we all needed to evacuate. But I was severely alarmed because as I was leaving…I could see that several pipes had cracked open, including what I believe were cold water supply pipes. That would mean that coolant couldn't get to the reactor core. If you can't sufficiently get the coolant to the core, it melts down. You don't have to be a nuclear scientist to figure that out." As he was heading to his car, he could see that the walls of Reactor 1…had started to collapse. "There were holes in them. In the first few minutes, no one was thinking about a tsunami. We were thinking about survival."

Another worker, a maintenance engineer who was at the Fukushima complex on the day of the disaster, recalls hissing, leaking pipes. "I personally saw pipes that had come apart, and I assume that there were many more that had been broken throughout the plant…I also saw that part of the wall of the turbine building for reactor one had come away. That crack might have affected the reactor….If the walls are too rigid, they can crack under the slightest pressure from inside…so it's designed to give in a crisis…That might be shocking to you, but to us it's common sense." A mile away a radiation alarm sounded at 3:29 pm, eight minutes before the tsunami hit.

In late March TEPCO went some way toward admitting at least some of these claims in a report titled "Reactor Core Status of Fukushima Daiichi Nuclear Power Station Unit One." The report said there was pre-tsunami damage to key facilities, including pipes.

Shaun Burnie, an independent nuclear waste consultant working with Greenpeace, observes, "Assurances from the industry in Japan and overseas that the reactors were robust [are] now blown apart. *It raises fundamental questions on all reactors in high seismic risk areas.*" Without exception, his assertion includes both nuclear power plants still operating in California: San Onofre and Diablo Canyon, both located on earthquake faults.

Meantime, largely unreported, just three hundred miles southwest of Fukushima, at a nuclear reactor perched on the slopes of a rustic peninsula, engineers are engaged in another precarious struggle. The Monju prototype fast-breeder reactor—site of a 1995 serious nuclear accident—has been in a precarious state of shutdown since a 3.3-ton device crashed and got stuck in the reactor's inner vessel, cutting off access to the plutonium and uranium fuel rods at its core. Engineers have been trying repeatedly since last August to recover the device. But the process is highly dangerous because the plant uses large quantities of highly flammable liquid sodium to cool the nuclear fuel.

65.
"Humankind Cannot Bear Very Much Reality"

August 18, the 160th day following the planetary disaster originating at Fukushima's defective series of GE Mark I reactors. Although I discover it only 22 days later, an unattributed communication appears in an internet video hosted by Thom Hartmann on Russia Today. "A lot of cracks come up in the ground. Massive steam is coming up from there. It's too smoggy here. Can't see a thing. It seems like nuclear reaction is happening underground. Now we are evacuating. Watch out for the direction of the wind."

Who is this Tokyo-based friend of Hartmann's who relayed this anonymous message to him? Who is the unidentified Fukushima employee who originated it? What is he describing here?

According to Hartmann's interviewee, Paul Gunter of Beyond Nuclear, readings at the site are one million millirems per hour. 500 rems is the lethal dose, but 1000 rems (twice the lethal dose) is coming out of these cracks, capable of causing fatalities within days.

The Japanese Nuclear Regulatory Commission's permissible radiation threshold for the public has been raised from 50 to 100 millirems per year. But this is one million millirems per hour. TEPCO has been looking to contain it by building a tent over the reactor, a remedy Paul Gunter opines will imperil the situation even further because it indicates TEPCO is planning to release radioactive steam in the high atmosphere, where it will spread even more.

About the time this begins to happen—and although I know very little yet of the reality of it—I have a telephone conversation with my son. I share with him my deepest fear that if there's an uncontrolled meltdown, if TEPCO can do nothing to stop it, molten fuel will reach the water table. My son asks me if we can't change the topic. And I do.

Twenty-one days later, I am packing for a short holiday—my first in two years. Before leaving, I make a quick check of my messages.

The same message appears again, subject line Fukushima: Possible China Syndrome, datelined August 18. It reads: *"A lot of cracks come up in the ground. Massive steam is coming up from there. It's too smoggy here. Can't see a thing. It seems like nuclear reaction is happening underground. Now we are evacuating. Watch out for the direction of the wind."*

My gut begins to churn. Later today I will hear—really hear—for the first time that cells very similar to brain cells live in the human gut, and they inter-communicate brain to gut, but I've known this truth for a very long time—through years of living in a state of severe anxiety laid down as a pattern from early childhood.

"Cecile, what would you do if your house caught fire?" my father asks.

"I would get out." I am two years old.

But this house—*this* house—I can't get out of. This house is mortal as my body is mortal. There is no leaving without death, my death of the body, world death of the planet.

The phone is ringing. I haven't finished packing. By now my ride must be less than ten minutes away. It's my friend from earlier days. Her voice tells me she's churning too. She's got reason: she's divorcing a husband who doesn't want to leave.

"Yes...," I say, "yes," but I'm seeing the steam bursting through the cracks, geysers of steam, of scalding water erupting from the earth. Searing temperatures, the air thick as tear gas, and I say "yes." And

I think, good God, how fortunate my friend is to be caught in the trammel of her own catastrophe. She knows nothing of this much greater one.

66.
Why every GE Mark I Boiling Water Reactor the World Over Needs To Be Shut Down

> *The Mark I containment has a proven track record of safety and reliability for over 40 years and there are 32 BWR (boiling water reactors) Mark I reactors operating as designed worldwide.*

So reads General Electric's PR statement—*dated March 16, 2011*—but the facts prove otherwise. At Fukushima, the GE Mark I design failed catastrophically when the outer containment building was neither large enough nor strong enough to withstand the hydrogen build up that caused three reactors to explode. Not only. Spent fuel cooling pools in the GE Mark I boiling water reactor are located five stories up, at the top of the reactor buildings, and if both onsite and offsite power fails, as happened following the megaquake in Japan, the reactor is left without any residual power to run the diesel back up cooling pumps.

As early as 1972, Dr. Stephen Hanauer, an Atomic Energy Commission safety official, recommended that the pressure suppression system of the GE Mark I be discontinued and any further designs not be accepted for construction permits. Although his boss, Joseph Hendrie, (later an NRC Commissioner) agreed with him, he denied Hanauer's recommendation on the grounds that it could well mean the end of the nuclear power industry in the United States. In 1976, three GE nuclear engineers publicly resigned their jobs, citing dangerous shortcomings in the GE Mark I design. Late one night, prior to his resignation, one of them, Dale Bridenbaugh spoke of his misgivings to his boss. "It's ten pounds of energy in a five-pound sack." His boss warned that scrapping the design would mean the end of GE's share of the nuclear industry.

In 1986, the year of Chernobyl, Harold Denton, then the NRC's top safety official, told an industry trade group that the "Mark I containment, especially being smaller with lower design pressure, in spite of the suppression pool, if you look at the WASH 1400 (the Rasmussen Report) safety study, you'll find something like a *90% probability* of that containment failing" (Nuclear Information and Resource Service. "Fact Sheet on Fukushima Nuclear Power Plant").

To protect the GE Mark I containment from total rupture, it was determined to modify the GE Mark I by designing and installing the "direct torus vent system" in all GE Mark I BWRs to vent any high-pressure buildup. The reinforced pipe installed in the torus was designed to release radioactive high-pressure steam generated in a severe accident by allowing the unfiltered release directly into the atmosphere. In other words, reactor operators could now expose the public and the environment to unknown amounts of harmful radiation in order to save GE's containment—and GE's bacon.

Of the 16 GE Mark I BWRs in other countries, some have already been decommissioned, but there are 23 GE Mark I BWRs operating in the United States, all of which share the same vulnerability as those at Fukushima, and none of which have been shut down. Their locations by state are:

Athens, Alabama, population 115,000
Baxley, Georgia, population 4,400
Cordoba, Illinois, population, 1,000
Moline, Illinois, population 47,000
Morris, Illinois, population 51,000
Palo, Iowa, population 700
Plymouth, Massachusetts, population 56,000
Monroe, Michigan, population 21,000
Monticello, Minnesota, population 13,000
Forked River, New Jersey, population 25,000

Hancock's Bridge, New Jersey, population 12,000
Scriba, New York, population 7,300
Southport, North Carolina, population 3,000
Brownville, Nebraska, population 150
Delta, Pennsylvania, population 700
Vernon, Vermont, population 2,000

Additionally there are 12 GE Mark II and Mark III reactors located in:

Clinton, Illinois, population 7,000
Marseilles, Illinois, population 9,000
St. Francisville, Louisiana, population 1,800
Scriba, New York, population 7,300
Perry, Ohio, population 1,300
Salem Township, Pennsylvania, population 7,000
Limerick, Pennsylvania, population 18,000
Richland, Washington, population 47,000

Although the populations of many of these towns and townships are small in themselves, some are located within 50-mile distances of highly populated areas, and twelve GE Mark I BWRs are located in densely populated states such as Illinois, Pennsylvania, New York and New Jersey. Even before the events of March 11, serious accidents involving loss of coolant were reported in Peach Bottom and Brown's Ferry GE Mark I reactors located in Delta, Pennsylvania and Athens, Alabama, respectively. So much for GE's "safe," pre-Fukushima track record.

On May 17, 2011 Donald Person adds this comment to GE's March 16 PR announcement: "The real flaw in the GE BWR Mark I design was the unfortunate assumption that no core melt would ever happen....You can erase this comment, suppress the details of this critical design weakness in the GE/Hitachi BWR reactor but you can't erase the facts as they will eventually come out elsewhere."

The question is, will the facts come out in time to prevent more planetary Fukushimas?

67.
What a Nuclear Disaster Originating in Japan Means for the Planet

On April 26, 1986, Chernobyl sent 190 tons of radioactive uranium and graphite into the atmosphere. It was described as the catastrophe of the century. The biggest disaster in the history of mankind. Three hundred forty thousand (340,000) troops were mobilized to deal with the disaster. Three thousand six hundred (3,600) Russian soldiers were used as human biots, ordered to pick up spent fuel with their bare hands. The casualty results from Chernobyl total one million deaths, 200,000 people incapacitated; tens of thousands of children born with atrocious birth defects. To this day, hundreds of Ukraine orphanages are overwhelmed with children born with physical deformities and mental handicaps. Demolition workers cleaning up Chernobyl received a fatal dose of radiation after only five minutes of exposure, and the 50-mile radius exclusion zone remains uninhabitable to this day (IAEA. "Report of the UN Chernobyl Forum Expert Group 'Environment' (EGE). 'Environmental Consequences of the Chernobyl Accident and Their Remediation: Twenty Years of Experience'").

On March 12, three GE Mark I BWRs at Fukushima, were on the verge of meltdown. Each GE Mark I contained 100 tons of fuel, fuel stored five stories above ground level. Unit 1's reactor suffered a complete meltdown. Hydrogen built up and exploded, rupturing containment and releasing both the core's fuel and eight years worth of spent fuel into the world's atmosphere. Two days later, according to Professor Edmund Lengfelder, Unit 3 may have suffered a hydrogen explosion, but there weapons-grade plutonium (MOX) was also being used as fuel. When the hydrogen gas exploded, it may have triggered a small nuclear explosion, comparable to a mini fission nuke (ENENEWS ADMIN. "German Radiation Professor Warns of Possible Nuclear Explosion at Fukushima").

Plutonium is the deadliest element known to humankind. Some sources claim that ten pounds (10 lbs) of plutonium evenly distributed could kill off the world's population ten times over. There are *10 tons* (that's 20,000 pounds) of plutonium at Fukushima. Although unit 4's reactor is empty, its spent fuel pool has eight years of spent fuel cooling. When power is lost, and water in the spent fuel pool boils off, it leaves spent fuel rods exposed, releasing hydrogen, which then explodes. In fact this may have happened. Pictures dating from December 5, 2011 indicate unit 4's south wall is missing (ENEWS ADMIN. "Report: "Confirmed that the wall of reactor 4 was lost on the south side").

In the first week after the earthquake, Fukushima has released more radioactive cesium than Chernobyl and all the bombs detonated during the years of atmospheric testing. One hundred (100) tons of fuel in the each of the cores of Units 1, 2 and 3, have melted through containment and fallen into the basement of the reactor buildings—something TEPCO admitted only much later. Thousands of tons of radioactive water have been released into the Pacific, contaminating water and sea life for all eternity or 4.5 billion years, whichever comes first.

Uranium has a half-life of billions of years. Entombing can't begin because radiation levels are so high that engineers can only work for a maximum of 5 minutes. Meantime, wind and weather patterns are spreading hot particles throughout the world. Citizens of Tokyo are inhaling 10 hot particles a day. In Seattle and much of the West Coast, people are inhaling 5 hot particles a day. Hot particles lodge in the lungs and bone, causing potential spikes of cancer and birth defects some years down the line. If we take Chernobyl as a benchmark, over time millions of deaths will result. According to recent CDC Morbidity and Mortality Weekly Reports, since March 11, infant mortality has risen 35% in 8 northwestern cities of the United States (Janette D. Sherman, M.D. and Joseph Mangano: "Is the Dramatic Increase in Baby Deaths in the U.S. a result of Fukushima Fallout?").

Radiation contamination has been detected in rain water, milk, fish, dairy products, vegetables, fruits and beef in the United States, although none of this has yet been reported by the EPA or by the "mainstream" media.

Millions of people in Japan are possibly living in radiation levels higher than the radioactive zone at Chernobyl. Thirty-four thousand (34,000) Japanese children between the ages of 4 and 15 are wearing radiation detectors to monitor the presence of iodine 131 in their thyroid glands. At the request of the Japanese government, the EPA has stopped monitoring fallout over the United States since May Before that it pegged iodine, cesium, xenon, and uranium levels at over 100 times the official limits.

If the Chernobyl disaster was rated a 7 by the IAEA, how do you think Fukushima should be rated?

68.
Birth Deformities

At the heart of the nuclear nightmare is something no one wants to talk about: birth deformities, a whitewash word for children born without the attributes we recognize as human. They are the hidden casualties of irradiation in the womb. Thousands of these beings draw breath in the darkened rooms and corridors of orphanages, state institutions, and in families brave enough—or poor enough—to have to keep them, these constant reminders of a man-made tragedy that can never be undone. These beings live in concealment in the Ukraine as the result of Chernobyl—and in Kosovo and Iraq, and wherever else NATO's occupying armies have irradiated the soil by dropping DU-tipped bunker busters and firing tank-piercing ordnance.

These beings pay the price and their guardians pay the price every day for nuclear accidents and of depleted uranium (DU) war. They pay the price for the accumulation of radioactive waste throughout the world over the past sixty years—in such enormous and horrifying quantities, governmental atomic agencies keep dreaming up ways of trying to make use of it, recycling nuclear waste whenever cancer patients receive radiation, whenever bone density tests are performed, whenever food—any food—is irradiated under the pretext of preserving it; in the tail assemblies and for ballast in Boeing 747s, and whenever DU is fused in the casings of ordnance of whatever caliber to be exploded, preferably in oil-rich countries, by the United States and its NATO surrogates. The winds and dust storms spread this radioactive dust throughout the world, ultimately irradiating the entire planet.

Almost every documentary film made of Chernobyl allows the viewer their full moment to recognize the image content, sufficient light to make it out—with one exception: birth deformities, beings born with horrifying malformations: three heads, no eyes, lifeless limbs twisting out of rag-doll torsos. These are the images mankind strives to keep itself from seeing—these, and images of death and dying. These are the dark secrets kept in the subterranean vaults buried beneath the bright and lying surfaces of living. (No other writer that I know of has dealt with buried horror more truthfully than Ursula LeGuin, in her short story titled "Those Who Walk Away from Omelos.") Yet every day, in our children's toys, these images abound: the face of swamp man; Mr. Potato Head (minus the popping Disney eyes that make it 'cute'). Picture these images now but with your own astonished eyes. Recognize in them the misshapen, grotesque beings that live in the hundreds of thousands, shut away in orphanages where none but the merciful with their forgiving eyes and hands manage to find it in their hearts to care for them.

How might a woman feel giving birth to a child with such deformities? Recent Bay Area newspaper reports describe one such woman, a woman accused of leading her four-year-old, severely handicapped child up four stories in a garage elevator and pushing him off the roof into the street below. He suffered from torticolis, a twisting of the neck, and a condition resulting in a severely misshapen head. To reporters, her weeping husband said his wife should not be held responsible. She did not want this child. After his birth she suffered severe postpartum depression. He said how much he loved her, and that he could not find it in his heart to blame her.

Now imagine appointing one woman whose tragedy it was to be caught pregnant, especially in her first trimester, when Chernobyl blew, or when Fukushima blew. Imagine appointing her to join the panel of engineers charged with designing the "new generation of reactors" trumpeted by the Obama Administration, or the board of the Nuclear Regulatory Commission, the NRC. Imagine appointing her sister to the International Atomic Energy Agency (the IAEA). What effect would it have on all those men sitting on their boards—the suited hijackers of our planet—to be met with these silent women, holding their atrociously deformed children in their arms? Imagine these women forcing that child into the arms of each board member to examine and caress.

What does it mean for me today, the one hundred sixty-fourth day following our planetary nuclear catastrophe, to plumb the darkened dungeon of my own imagination, the place I have resisted from the beginning, to reach its darkest heart? What do I feel, I, a young-bearing animal, a woman, a mother? At my deepest heart, I feel rage boiling up, a cyclone which, if I let it, would explode, vaporize the walls that contain me. How can I possibly meet this rage without dying of it? How can I stop it from tearing me apart?

69.
Earthquake Warning

On the one hundred sixty-fifth day following our planetary nuclear catastrophe, an earthquake of 5.9 magnitude rocks Virginia; its waves are felt in Washington D.C. and throughout the East Coast from North Carolina to Toronto. Two spires of the Washington Cathedral fall off; a building in Baltimore collapses; the Washington Monument shows severe cracks in at least four places. The White House and the Pentagon are evacuated. Terrified workers in New York City's Water Street hit the street. Two reactors at the North Anna Plant operated by Dominion Virginia Power shut down automatically despite the immediate failure of one of the backup diesel generators. Designed to withstand a 6.1 earthquake, they are located directly over a fault.

Willie Martin, now "80-something," worked with the contractor that built the North Anna plants in the 1970s. Martin said experts came out to show the construction workers where the fault line ran. "They said it was a fault, but the ditch looked pretty much the same on both sides to me."

Six other reactors reported unusual events. El Presidente plays golf in Martha's Vineyard. The "deciders" are on their August congressional break, most of them too far away to feel it—assuming they could feel anything at all.

Both New York State's Indian Point, located one mile from an earthquake fault, and Pilgrim, near Boston, lack designs appropriate to withstand the kinds of earthquakes that might occur there, putting millions of people in mortal danger. At least three earthquakes located on the Hayward Fault in the 2 and 3 magnitudes have shaken California within the last 24 hours—California, where two reactors, San Onofre and Diablo Canyon, are located over major faults.

70.
Somewhere Else on the Planet...

August 29, one hundred and seventy-one days following our planetary nuclear catastrophe, more articles appear citing Chernobyl—or Hiroshima—as benchmarks to assess the magnitude of Fukushima's disaster. The August 2 readings of 10,000 millisieverts per hour detected at the plant—equal to some 100,000 chest X-rays—used equipment which could only measure at a distance, incapable of gauging the exact level, because the device's maximum reading is only 10,000mSv. Despite this, and despite the fact that radioactive cesium in processed tea has been detected as far away as 160 kilometers from Fukushima, there seems to be no public readiness to contradict the Japanese government's decision that the uninhabitable zone at Fukushima stretch no more than the 20-kilometer radius originally declared. There is widespread concern in Japan about a general lack of government monitoring, and people who decide to do their own independent monitoring—and their own do-it-yourself decontaminating—are finding disturbingly high levels of radiation

Dr Tatsuhiko Kodama, a professor at the Research Centre for Advanced Science and Technology, and Director of the University of Tokyo's Radioisotope Centre, believes things are far worse even than indicated by the recent detection of extremely high radiation levels at the plant. In a July 27 speech to the Committee of Health, Labor and Welfare at Japan's House of Representatives, he claimed that "the government and TEPCO have not reported the total amount of the released radioactivity yet." Taking readings from 27 facilities measuring radiation across the country, Kodama's center has discovered that "the total amount of radiation released over a period of more than five months from the ongoing Fukushima nuclear disaster is the equivalent of more than 29 Hiroshima-type atomic bombs," and the amount of uranium released "is equivalent to 20 Hiroshima bombs." Although he made the statement in July, it only became public knowledge in late August.

Dr. Yuko Yanagisawa, a physician in neighboring Chiba prefecture, reports that she's beginning to see health effects in children indicative of radiation disease: increased nosebleeds, unrelenting diarrhea, and flu-like symptoms. In her opinion, all the government has done so far is raise the acceptable radiation exposure limit for children from 1 mSv/year to 20 mSv/year—which it did in April. Nonetheless, in the period between March 24 and March 30, medical tests done on children living in three towns near the plant found 45% were already suffering low-level thyroid radiation exposure.

Dr. Helen Caldicott, founding president of Physicians for Social Responsibility, observes that new exposure standards by the Japanese government don't take differences between children and adults into account. Children's susceptibility to radiation exposure is several times higher than that of adults, and women's susceptibility is higher than men's. Speaking to Al Jazeera, she stated further, "Radioactive elements get into the testicles and ovaries... [causing] genetic diseases like diabetes, cystic fibrosis, and mental retardation...[which] are passed from generation to generation, *forever*" [italics mine] (*Al Jazeera English*. "Fukushima Radiation Alarms Doctors").

Yet only this summer, the UK-based Economist Intelligence Unit (EIU), a decades-old economic research group, found that the impact of the Fukushima disaster, now widely considered the worst industrial accident in history, "is expected to be minimal." According to the EIU, Germany's decision to scrap eight nuclear power plants and place its remaining nine under review will be outstripped by China, India and Russia, which will add five times the nuclear capacity that the German decision removes from world nuclear output. In fact, sixteen new reactors began construction in 2010—ten in China, with others in Russia, India and Brazil. Even Japan has resumed construction work at a new nuclear plant, after putting additional anti-earthquake measures in place. Indeed, the report points out, by 2015 one new nuclear reactor will come online every month somewhere on the planet. And in Japan, although recently-resigned PM Kan called for Japan to end its reliance on atomic power after the world's worst nuclear accident in 25 years, his successor, PM Yoshihiko Noda has a different message for the people of Japan, still reeling from the shock of the world's worst nuclear accident: atomic power is needed to save the Japanese economy.

Given its centuries-old culture, and its past history of deep distrust of Russia, it's a long shot that Japan will respond favorably to the recent suggestion by an American woman that the government of Japan buy territory in Russia equivalent to Japan's, where the Japanese people would be allowed to re-create their nation in a less seismically active zone.

71.
And Bringing Up The PR Rear...

On the one hundred seventy-second day of our fallout world, Exchange Monitor Publications & Forums announces THE FIFTH ANNUAL—yes, here it is ladies and gentlemen!—the FIFTH ANNUAL RADWASTE SUMMIT you've all been waiting for, a chance to meet and greet and booze and schmooze in Las Vegas' culture capital!!! And with accommodations at the Marriott, where you'll be ENTERING A NEW ERA OF RADIOACTIVE WASTE MANAGEMENT! while enjoying drinks at poolside at our GORGEOUS SUMMERELIN RESORT AND SPA!!!!!

Chew the fat with David Martin, Secretary, New Mexico Environment Department—you remember Dave! And Amanda Smith, Director, Utah Department of Environmental Quality, can't beat Mandy! And Frank Marcinowski, Deputy Assistant Secretary for Technical & Regulatory Support, DOE-EM, good ol' Frank, and Gerard Bruno, Head, Radioactive Waste and Spent Fuel Management Unit, IAEA—that's Gerry to you—and Larry "good 'ol boy" Camper, Director, Division of Waste Management & Environmental

Protection, U.S. NRC, and Sharron DaCosta-Chisley, National Program Manager, U.S. Army Corps of Engineers—hey! Sharon don't pull no punches! And Ralph Andersen, Senior Director, Radiation Safety & Environmental Protection Nuclear Energy Institute (we got him here just for the record), and Alan Parker, President, Government Group Energy Solutions—he's got all the answers I know you all wanna hear, and Rod "life of the party" Baltzer, President, Waste Control Specialists...and many more....

Our agenda kicks off with our very own NRC Commissioner Bill Magwood—and I know you all will be wanting to hear what Bill has to say—before we meet downstairs afterwards for drinks and your chance to schmooze with him. You'll be sharing the action with all the folks from the industry—many of them repeaters. And we got a drop-dead guest this year who made it all the way from the Ukraine—that's the USSR to you—Tatiana Kilochytska, to talk to us all about what the folks out in the Ukraine did about waste management when things got hella hot!

72.

Killing A Nation/ Killing A Planet

Connecting some dots: nuclear energy produces radioactive waste, mostly in the form of spent fuel rods; storage of radioactive waste in cooling pools poses yet more danger than the reactors themselves, because the pools are uncontained and unprotected, and often lack adequate emergency back-up systems. Deep storage of nuclear waste is problematic: the Swedo-Finnish Onkalo is not equal to providing storage, probably not even for the limited 100,000 years for which it is designed. The Yucca Mountain project, the U.S.-designated dump located in Nevada, was only recently zeroed out; it presented porosity problems such that storage casks corroded on contact with steady leaching of water. Civil applications provide very limited avenues for disposal. Other "waste" products include DU, the by-product of U-235 enrichment. Only one problem remains to be worked out: DU's effects are irreversible soil, water, and air contamination far beyond its half life of 4.5 billion years, resulting in birth deformities and cancers for all eternity. Is it conceivable we might expect to see cover-ups in connection with its use?

73.
Everything You May Not Have Wanted To Know About DU And Preferred Not To Ask

Depleted uranium (DU) is an alloy of the radioactive, heavy metal uranium, with a lower content of the fissile isotope U-235 than naturally-occurring uranium. It is a byproduct of the uranium enrichment process that isolates the fissile U-235 isotope for use, both in nuclear bombs *and* in nuclear reactors as fuel. It has a typical isotopic content of 99.75% U-238 by weight, 0.25% U-235, and 0.005% U-234 with a half life of 4.5 billion, 700 million and 245,500 years respectively (Wikipedia. "Depleted uranium").

Its highly ignitable properties allow the DU penetrator to ignite on impact, generating extremely high temperatures. As it pierces its target, it leaves burned-off material behind, dispersing DU dust into the environment. The quantity of aerosol production is proportional to the DU mass within the projectile and the strength of the impact. Up to 70% of DU contained in the projectiles is aerosolized on impact when the DU catches fire. The explosion generates temperatures of between 3,000° and 6,000° Celsius. Nanoparticles so generated are smaller than 10 microns in size and act more like a gas than a particle. Because they remain air and windborne for long periods of time, they constitute the most serious threat to any human population close to battle areas. Marion Fulk, a retired chemical physicist who built nuclear bombs for more than 20 years at Lawrence Livermore Lab, identified DU weapons as "dirty bombs in every way." (Dipnote. Official blog of the U.S. Dedpartment of State).

Depleted uranium was first manufactured in the United States in 1968. In 1972, it was given to Israel to use in the Yom Kippur War. Later it was used in warfare in Somalia, in Kosovo and in Bosnia; it was used in the "Highway of Death" during the first Gulf War in the form of DU-tipped tank-piercing ordnance, destroying all of Saddam Hussein's retreating army. It was used again in Afghanistan and during the ten-year no-fly zone sanction period, in which some one million Iraqis died, 500,000 of them children; it was used in "Operation Iraqi Freedom," otherwise known as the shock and awe carpet bombing of Baghdad. Today it probably is being used in Yemen, Somalia, and Libya as well.

DU is used by both the US and UK military for four main reasons: 1) it's cheap and in plentiful supply; 2) it's effective in military engagement because its high density and self-sharpening qualities enable it to penetrate hard targets easily; 3) it's pyrophoric, that is, it burns on impact, increasing its ability to destroy enemy targets; and 4) it makes use of the piled up by-product of U-235 enrichment.

It is used in Abrams tanks, A-10 warplanes, and Bradley fighting vehicles, in bombshells, tank armor plating, aircraft ballast, Bunker Busters, and fragmentation mines (otherwise referred to as "anti-personnel devices"). At least 15 other countries hold supplies of DU.

In all, 320 tons were used in the 1991 Gulf War; an appalling 90 tons were used in only the first 24 hours of the "shock and awe" carpet bombing of Baghdad when uranium aerosols were widely dispersed in the atmosphere and blown across Europe so that 9 days later, on March 28, elevated levels of uranium exceeding Environmental Agency thresholds were picked up in various areas in Great Britain, a clear-cut case of the chickens coming home to roost.

74.
Where You Might Have Been And When

Exhale. Inhale. Depending on where you were and when you were there, you have just inhaled a nanoparticle of DU (depleted uranium) smaller than 10 microns. If you served in the First Gulf War, particularly around Basra in southern Iraq, and particularly if you found yourself near the "Highway of Death," where Saddam's tanks and personnel carriers tried making a dash of it out of Kuwait and back toward Basra, you most certainly inhaled many nanoparticles of DU. You came home from the three-week war with particles lodged in your lungs, particles that passed through your alveoli and into your bloodstream, eventually to lodge in your muscles, in your bones, in your brain, wherever the red rivers of your body took them, and you became a statistic. If you were assigned to clean out damaged tanks, tanks caught in crossfire, you will most certainly have inhaled DU. If you were assigned to prepare damaged tanks and personnel carriers for shipment and burial in the United States as the government of Kuwait demanded after the war, you are sure to have inhaled DU, or it may just as easily have penetrated through your skin. You were never warned by any commanding officer that you would come into contact with DU. Sixty-five of a hundred of you would have children with severe illnesses, missing eyes, blood infections, respiratory problems and fused fingers, because DU remained lodged in the place it came to rest within your body, irradiating neighboring cells in a kind of 'bystander' effect, leading to an 8-fold increase in genetic damage. You are three times more likely to have children with birth deformities than fathers who did not serve in the Gulf. Your babies are 65% more likely to have physical abnormalities, and your wives and partners are 40% more likely to miscarry. They are the lucky ones.

By 2003, following the First Gulf War, Basra saw a five-fold increase in congenital malformations, and a quadrupling incidence of malignant diseases. If you were a child of Basra, playing in abandoned vehicles or tanks, you would have inhaled DU and probably ingested it. You, too, would become a statistic—an Iraqi one—one of hundreds of thousands of Iraqi children developing childhood leukemia. You would be five times more at risk for developing cancer of the thyroid. You would be condemned to a short life of bodily suffering, and while you sickened and died, you would have to see the pain in your mother's eyes, your father's grief—if you still had a mother or a father. If you were or are a woman in Basra, your risk of giving birth to a severely handicapped child would have increased 60%. Either you might have been helpless to prevent a pregnancy, or perhaps you would have ignored the advice of those Iraqi doctors who had not already fled Iraq who said, "Iraq is no longer a place to have children," because had you given birth to such a child, you would have had to care for it with the little food available to you, bathed him with contaminated rain or sewer water. Perhaps you would have had to carry him, if in his short life he could never have hoped to walk, or see, or feed himself—or any of the kinds of things that characterize human living in the world. And you yourself might be one of the 70 out of 1,000 people to develop cancer (*Viewzone*. "Depleted Uranium: Anatomy of an Atrocity").

Or suppose that you had become pregnant in Falluja following the assaults of April or November 2004, in which NATO used white phosphorous and DU. According to the *International Journal of Environmental Research and Public Health,* your chances of giving birth to a child with neural tube defects affecting the brain or lower extremities, with cardiac or skeletal abnormalities, or with cancers, would have multiplied eleven times, and your city would have seen a 15% drop in the number of boy children born. In a study conducted by Dr. Samira Ghani, 11% of babies born in Fallujah were premature (under 30 weeks), 14% of fetuses spontaneously aborted, and of 547 babies born to a sampling of 55 families, 15% had serious birth defects. Such defects, which were extremely rare in Iraq prior to the large-scale use of DU weapons, are now commonplace (Martin Chulov. "Research links rise in Falluja birth defects and cancers to US assault"). (Photographic documentation showing such anomalies as infants born without brains or faces, with their internal organs outside their bodies, without sexual organs, or without spines, can be seen at http://stgvisie.home.xs4all.nl/VISIE/extremedeformities.html.)

Their cover-ups notwithstanding, the U.S. military is aware of DU's harmful effects on the human genetic code. Already in 2001 Dr. Alexandra Miller had completed a study for the Armed Forces Radiobiology Research Institute at Bethesda indicating that DU's chemical instability causes one million times more genetic damage to DNA than would be expected from its radiation effect alone. But by interfering with their rights to conduct their own investigations, the U.S. has prevented Iraqis from obtaining the data needed to confront the enormity of the health damage to the population caused by DU (Prof. Souad N. Al-Azzawi. "The Responsibility of the US in Contaminating Iraq").

Dr. Hardan is a special advisor to the World Health Organization. He documented the effects of DU in Iraq between 1991 and 2002. "American forces admit to using over 300 tons of DU weapons in

1991. The actual figure is closer to 800. This has caused a health crisis that has affected almost a third of a million people. As if that were not enough, America went on to use 200 tons more in Baghdad during the recent invasion." According to Dr. Ahmad Hardan, writing about Basra, it took two years of research to obtain conclusive proof of what DU actually does:

> . . . but now we know what to look for, and the results are terrifying. Nothing can prepare someone for the sight of hundreds of preserved fetuses scarcely human in appearance, or babies born with terribly foreshortened limbs, with intestines outside their bodies, with huge bulging tumors where their eyes should be, with single eyes, like a cyclops, without eyes, or limbs, and even without heads. Such defects are almost unknown except in babies born near A-bomb test sites in the Pacific.

Dr. Hardan reports that he arranged for a delegation from Japan's Hiroshima Hospital to visit in order to share their experience about the radiological diseases Iraq was likely to see over time. But this delegation declined to come after the Americans objected. Likewise, a world-ranking German cancer specialist agreed to come, only to be denied entry into Iraq.

John Hanchette, who is a professor of journalism and one of the founding editors of *USA Today,* reports that, although he prepared news breaking stories about the effects of DU on Gulf War veterans and Iraqi citizens, each time he was ready to go to press, he received a phone call from the Pentagon requesting he not publish. Since, he has been replaced as editor of *USA Today.*

Dr. Keith Braverstock, the World Health Organization's chief expert on radiation, alleges that his report "On the Cancer Risks to Civilians in Iraq from Breathing Uranium-Contaminated Dust" was deliberately suppressed.

Dr. Alim Abdul Hameed Yacoub published three studies between the years 1998 and 2002 indicating that malignant diseases in the children of Basra showed a 60% rise from 1990 to 1997, including a 120% increase in malignancies in all children under the age of 15. He demonstrated a 160% increase in uterine cancer, a 143% increase in thyroid cancer, a 102% increase in breast cancer, and an 82% increase in lymphomas. His results also showed that new generations of cruise missiles contain DU.

Dr. Yacoub was twice attacked in his home by pro-occupation militias two weeks before he was killed, along with his son, when his car was forced off the highway on his way back home to Basra. Other researchers were deprived of their freedom by imprisonment without accusation, and some 250 Iraqi scientists were assassinated after Iraq's invasion by occupation militias (Dr. Souad Al-Azzawi: "Depleted Uranium Radioactive Contamination in Iraq: an Overview").

Dr. Asaf Durakovic, Uranium Medical Research Center (UMRC) founder, accompanied by a research team, conducted a three-week trip in October of 2003, collecting some one hundred samples of soils, civilian urine and tissue from Iraqi soldiers in ten Iraqi cities. They determined that their samples revealed hundreds of thousands of times normal radiation levels. When he went on record stating that the U.S. doesn't "want to admit that they committed war crimes by using weapons...banned under international law," he was warned to stop his work. Subsequently he was fired from his position, his house was ransacked, and he reported receiving death threats.

The Pentagon and the Department of Defense have interfered with publication of UMRC's studies by a campaign of misinformation in the press, through the funding of research grants refuting UMRC's scientific findings, and by destroying the reputation of UMRC's staff, physicians and laboratories. The re-framing of the data as a debate between government and experts polarizes the issues and obscures the scientific facts based on irrefutable data. In this, the UN's regulatory agencies, the military, and defense sector manufacturers are all complicit (Doug Westerman. "Depleted Uranium—Far Worse Than 9/11").

75.

Poison Factories, Poisoned Earth

Who are these manufacturers of DU, and where are their plants located? A partial list includes:

- Starmet Corp. (formerly Nuclear Metals) in Concord, Massachusetts.

- Oak Ridge Centers for Manufacturing Technology (and Y12) in Oak Ridge, Tennessee.

- Rocky Flats Environmental Technology Site (formerly Rocky Flats Nuclear Weapons Plant) near Boulder, Colorado.

- Fernald Environmental Management Project (formerly Feed Materials Production Center) in Fernald, Ohio (until 1989).

Present and past manufacturers of DU ammunition include (among others):

- Starmet Corp. (formerly Nuclear Metals) in Concord, Massachusetts.

- Primex Technologies in St. Petersburg, Florida

- Alliant Ammunition and Powder Co. in Radford, Virginia.

- Radford Army Ammunition Plant in Radford, Virginia.

- Aerojet Ordnance Co. (formerly Aerojet Heavy Metals Co.) in Jonesborough, Tennessee.

- Twin Cities Army Ammunition Plant in New Brighton, Minnesota.

- US Army Materials Technology Laboratory in Watertown, Massachusetts (until 1995).

• Lake City Ammunition Plant in Blue Springs, Missouri (until 1985).

• National Lead Industries in Colonie, New York (until 1980).

In 1979, DU particles escaped from the National Lead Industries factory at Colonie, which was turning out DU weaponry for the U.S. military. The factory released more than .85 pounds of DU dust into the atmosphere every month, prompting a cleanup cost of contaminated surrounding property in excess of $100 million dollars before the plant closed in 1980 (Vladimir Zajic. "Review of Radioactivity, Military Use and Health Effects of Depleted Uranium").

Past disposal practices at DU manufacturing sites have contaminated sewers, soil, groundwater, and municipal water supply with DU and other harmful substances. Cleanup costs at the Twin Cities Army Ammunition Plant alone were estimated at $230 million.

And where are the test sites where DU weaponry is tested? A partial list includes:

• Yuma Proving Ground, Arizona. Soil surface is contaminated with DU, 1500 times normal background levels. Samples of wildlife show DU contamination. 25% of kangaroo rat kidney samples exceed the toxicity threshold for uranium. (The Western Shoshone rely on kangaroo rat kidneys for food.)

• Eglin Air Force Base, Florida. 220,000 pounds of DU penetrators have contaminated sand, which is periodically shipped and deposited to the low-level dump at Barnwell, S.C. 90-95% of the uranium remains on site.

• Nellis Air Force Base, near Las Vegas, Nevada. 77,000 tons of 30mm DU munitions are stored at NAFB in addition to nearly 200 tanks and vehicles that have been fired upon with DU munitions. The DU licensed area is so hot, it's off-limits; it can be accessed only once or twice a year for cleanup.

• White Sands Missile Range, New Mexico. Several earth-penetrating missiles, each using cores of 80 kg of DU, were tested. One reached a depth of 200 ft, such that the core could not be recovered. Its shaft is filled with anywhere from 70 to 100 ft of ground water showing concentrations of uranium ten times the background level. — http://www.wsmr.army.mil/

• Los Alamos National Laboratory, New Mexico. 220,000 pounds of DU were expended since the beginning of operations. 90% remained around the firing ranges, and 10% has already entered the watershed.

• New Mexico Institute of Mining and Technology, New Mexico obtained a 99-year permit for open testing of DU munitions. Their application described the test area as so contaminated with DU as to preclude any other use.

• China Lake Naval Air Warfare Center (formerly China Lake Naval Weapons Center), California contains some 25,000 pounds of DU fragments. Lockheed got a $4 million contract to separate DU from the soil.

• Lake City Ammunition Plant in Missouri. When the 20mm M10-1DU penetrator was declared obsolete, 44,000 rounds were destroyed—some 7,700 pounds. Some of the DU fragments were shipped to a waste disposal site, but 11,000 cubic meters of soil is still contaminated.

• Ethan Allen Firing Range in Vermont. After 200,000 anti-tank rounds containing 10,000 pounds of DU were discharged, 4 inches of contaminated topsoil had to be scraped and shipped to a low level radioactive dump.

• Vieques Island, Puerto Rico is a special case. Appropriated by the Navy and Marine Corps. in WWII, it was used as a toxic waste storage site and a bombing and gunnery range for over 55 years. Repeated protests by Viequeños citing heightened cancer incidence and birth defects had no effect, until a civilian guard was killed by two misfired bombs. Citizens invaded the proving grounds, doctors joined them citing the Hippocratic oath; fishermen sailed their boats dangerously close to Naval ships, all of which attracted enough media attention that the Navy finally left but refused to clean up. The island was officially designated by the U.S. government as a Nature Preserve.

• Aberdeen Proving Ground, Maryland. Contaminated with 154,000 pounds of DU penetrators. Soil samples show anywhere from 16,000 to 50,000 times background level, suggesting that DU was redistributed by dissolution and transport with water, with the highest concentration of DU in zooplankton. From there, DU will eventually enter the human food chain, although most DU would settle in sediments after a period of a 100 years.

• Jefferson Proving Ground in Indiana. 23 million shells and bombs and 152,000 pounds of DU penetrators were fired. The facility closed in 1995. Clean up costs are in excess of 4 to 5 billion dollars. Soil DU contamination ranging from 20 to 60 cm averages 15 cm deep. Excess radioactive exposure from drinking groundwater by humans would increase after a 30-year time lag because DU will leach into the aquifer, reaching a level of 20 mrem/year two centuries after contamination, tapering off to background levels in another eight centuries.

In all, 42 states are contaminated today with DU from manufacture, testing and deployment. Women living around these facilities have reported increases in endometriosis, birth defects in babies, leukemia in children and cancers and other diseases in adults.

But on the one hundred and eighty-first day following our planetary catastrophe at Fukushima, the millions assessed for clean-up tell us little or nothing about the incidence of cancer near U.S.

domestic sites; they tell us nothing about the spike in U.S. birth defects, nor about spontaneously aborted fetuses. They tell us nothing about the Home Battlefield here in the U.S. where DU is manufactured, transported, and tested before export to foreign battlefields throughout the world.

76.
A Global Invitation

The e-mail invitation stares me in the face. One hundred thousand poets are scheduled to participate in support of one million marchers world-wide protesting the global corporate take-over. There's a response window licking at me like a hot temptation. I toss in my own misery:

One hundred thousand poets, one million marchers. How can all these good sentiments stop the juggernaut of the corporate, international terror state? Our level of discourse is words, their discourse is depleted uranium tipped bullets. Depleted uranium (DU) is another way of saying "dirty bomb." Three hundred fifty areas of Iraq soil are now contaminated FOR ALL ETERNITY. Localized contamination will continue spreading by dust and wind storms, sandstorms and rainstorms and will be spread FOR ALL ETERNITY. Fifteen in 100 Iraqi babies are born horrifically deformed because the chromosomes of their parents have been damaged by DU exposure. One in three Iraqis has cancer. The Pentagon prevents anyone reporting on these statistics through a program of censorship, disinformation and dirty tricks. That is why "immunity" is in the pipeline for all U.S./NATO troops deployed there lest they be charged with crimes against humanity.

Returning U.S. troops father severely deformed children; their urine tests positive for uranium from exposure to DU. There is a "reason" for all this: the Pentagon's policy with respect to depleted uranium, namely U-238, the by-product of U-235 enrichment, of giving it away practically free of charge to weapons manufacturers to tip ordnance for use by U.S. forces and their NATO surrogates, especially in oil-rich countries. But every time we turn on a light, or turn on our computers to compose our poems, we are benefiting from nuclear power; we are living in the pipeline that deliberately spews nuclear by-products on the soils of "other" people, members, like us, of the same human race. Let us remember that we are one human flesh. Let us make "words" that stop the murder of a planet, which, last time I looked, was not the property of General Dynamics, General Electric, or any other general murderer.

But what words will those be?

77.
How Prolific Does a Prolific Criminal Have to Be?

On the 188th day of our planetary disaster, Bonnie Urfer, a long-time nuclear weapons resister, is sentenced by a judge who pegs her as a "prolific criminal," an honor done her, much as Nehru saw his own imprisonment by the British in India's anti-colonialist struggle of the 40s. How prolific does a criminal have to be to win such an award?

Bonnie Urfer has worked for *Nukewatch*, the nuclear protest quarterly, for 25 years. She's done more than four year's time for peaceful non-violent resistance. Her offense? She crossed on to the property of the Y12 Nuclear Weapons Complex in Oakridge, Tennessee to protest the continuing production of thermonuclear weapons of mass destruction at Y12, a Class A Misdemeanor Trespass for which she received the maximum sentence.

Bonnie Urfer knows something about the many chapters in the nuclear play book: the toxic consequences to the body resulting from uranium mining; the decommissioning of spent fuel; the storage of spent fuel in unprotected cooling pools; the impossibility of safely storing the nation's growing stockpile of radioactive waste, and the policy of the government to pass the by-products of Uranium-235 enrichment to corporations manufacturing DU and DU-fortified ordnance at almost no cost. She is persuaded that as long as nuclear weapons are manufactured, they will be used. (Said Harry Truman of the bombs produced by the Manhattan Project: "They cost $6,000,000. We gotta use 'em.")

Said the sentencing judge: "The court finds you are a prolific criminal.... You are not deterred by sentences, and you have regularly informed courts that you will not comply in terms of supervised release and your actions have validated that promise. You most likely will not stop. So the only way to insure you will not continue this behavior that has become your routine...is to separate you from the community."

Judge Guyton sentenced Bonnie Urfer to eight months—the maximum sentence—less four months, three days for time served, a sentence meant to serve "as an adequate deterrent and a just punishment," and a reflection that Bonnie Urfer dared to speak truth to power. In her sentencing statement titled "So Many Crimes, So Little Time," Bonnie wrote: "It doesn't matter what my sentence is. If I am returned to jail, I'll expose more crimes. If I am set free, I'll expose more crimes....In the past 126 days I have been booked into three different jails. The hardest part...is being just one person in the midst of so much systematic crime. [Jails are] places with some of the worst human rights violations in one of the most unjust systems. It is important to know what happens in them:

• the on-going crime of medical negligence in these jails.

• the illegal starvation diet in the Blount County jail, for which no one has been arrested.

• the practice of overcharging mostly dirt poor inmates for phone calls, and commissary, so that corporations and counties receive greater kickbacks.

• the presence in this courthouse of those who show up protesting unjust sentences for nonviolent conspiracy charges.

• the prosecutors, judges, attorneys, court clerks and law enforcement personnel who hold stock in the private prison industry, commissary companies, phone providers or medical contractors in these human warehouses.

"I heartily disagree with this court that Y12's production of nuclear bombs does not equate to imminent nuclear war. I can tell you about the women I met in the jails who lost family members from cancer after exposure to radiation while working at Y12. The government pays $150,000 to those with cancer or to their family after a death if they can prove Y12's liability. Thousands of people are dead or dying from weapons production. How many deaths does it take to convince the courts that Y12 is killing its own in a nuclear war? How many does it take to name it a crime?"

78.
Two Halves Make One Whole…

I would never have learned about Bonnie Urfer without having read the text of her statement to Judge Guyton in Common Dreams. And I could not have quoted the judge's sentence had I not sat next to Megan Rice at breakfast.

I spend the one hundred and eighty-eighth day following our planetary catastrophe fitting two pieces of the same puzzle together in the company of my recently re-discovered, 80-year-old parochial schoolmate, Megan Rice, who in her early years took religious vows and who's just completed a four year stint in Nevada protesting nuclear war, and the use of drones and drone warfare at Creech Air Force Base and at the Nuclear Test Site, an area the size of Connecticut, appropriated by the U.S. Government from Western Shoshone lands to poison them with radioactive waste. We found we had been schoolmates—way back seventy years ago—when it dawned on us we'd both been to school with George Carlin.

She's lived a long life, at least 30 years served in Nigeria and elsewhere in Africa, teaching math and science, this woman who required an 8-month period of recollection on re-entry into the culture people—Americans—assume is the only way to be in the world. "Slow down," she admonishes, "slow down," like my Nigerian taxi driver exclaims after avoiding a nasty accident: "You are too much in a hurry."

She's here to say her good-byes before returning East. Today she is surrounded by women her age and much younger—all independent, all doing the work in the world each one is meant to do—which is why none talks about retirement. Each in her way is a justice worker, dedicated one way or another to making things right—even facing jail time if necessary—as my friend has done, her many arrests protesting the militarism and the culture of militarism that has the country by the throat.

To sit with them at lunch, to hear their laughter, to see their eyes light up and shine—it's not even in the words they exchange so much as in the kind of sound-like-music people make who are consciously engaged in the world and who have found their authentic place in it.

79.

Killing Our Own: Mr. Lindsay Goes Home to Die

On the one hundred eighty-seventh day following our planetary disaster, Georgia WAND (Women's Action For New Directions) reports being particularly concerned with the deteriorating health of communities located near nuclear facilities. Their concern is echoed in Joseph Trento's book, *The Bomb Plant,* which describes conditions at one of the U.S. Government's most secretive institutions, the National Nuclear Security Administration (NNSA) and its radioactive weapons facility, the Savanna River Site (SRS) where the world's surplus plutonium is stored in a quaint, minimally-secured old building on top of the South's most dangerous earthquake fault.

He describes the Lindsay family, a black family WAND has reached out to assist. Mr. Lindsay had ten children to support. He left his job as a segregated school principal to commute several hours a day to work at the SRS. Like thousands of other African Americans, he was given the most dangerous jobs and ordered to throw his dosimeter (his radiation exposure gauge) in a bucket before entering high-risk areas. He began to get sick. He brought radiation home on his clothing, exposing his children as they played with him or sat on his lap, exposing the wife who dutifully washed his clothes. His family even thought they were lucky because SRS management encouraged their African American employees to hunt, fish and pick fruit from the old orchards left standing after the five towns were lost to build the bomb plant. "Finally, one awful day," writes John Trento, "Mr. Lindsay was driven home to die. The radiation he brought home would reach from his grave to take wife, then child after child (Deborah Dupré. "Obama's Dirtiest Secret Exposed").

In the same report, John Trento describes how a MOX (weapons grade plutonium) plant is even now being built at SRS, America's most radioactive Superfund site, located on South Carolina's most dangerous fault line (John Trento. "MOX Fuel Rods Used in Japanese Reactor").

80.

Race and the Arms Race

Repeated contamination of "sacrifice zones" by the United States (and by the former Soviet Union) for weapons production, weapons testing, mining, milling, and enrichment, has primarily exposed people of color and otherwise marginalized people to unacceptable risks, consistently without any informed consent. With the advent of the Fukushima planetary-wide accident, directly attributable to GE's criminal concealment of the Mark I's design defects, and to TEPCO'S negligence, the population of the entire planet now joins the ranks of marginalized people such as the Navajo, the Western Shoshone, Inupiat, Marshall Islanders, Yanomami of the Peruvian Amazon, the Moslem stockmen and Russian peasants of Chelyabinsk, the Kazakh herders of Kazakhstan, the Black population of South Carolina, the Mormon stockmen of Utah, some of whose stories appear below:

Diné (Navajo) Nation
Mining and Milling in New Mexico

Some 270,000 Navajo (Diné as they prefer to be called) lived in 26,000 square miles of the Southwest. Early in the Cold War, in the rush to obtain uranium, the U.S. Government awarded contracts to mining and milling companies, even though the uranium deposits were actually located on lands belonging to the Navajo Nation. They sold the idea to the tribal council by promising them jobs.

The Diné worked in mines lacking ventilation; they ate their lunch inside the tunnels, drank water from water sources inside them. No toilets, showers, change houses, or drinking water were provided them by the mining companies. They returned home with radioactive clothing. Although the AEC controlled all the stockpiles of uranium, they disclaimed responsibility for miner health. The U.S. Public Health Service actually colluded with mining corporations to obtain miner's names in exchange for not advising them of the dangers associated with uranium mining and providing no information about how their subsequent illnesses may have been associated with their exposure to uranium. When the Diné began to suffer severe health effects, they had no word in their language for radiation, no word for radiogenic illness. They had no way of understanding what the causes might be, and even if they had, their access to health and legal services was very tenuous. One Navajo widow whose miner husband died of cancer lived in a remote area. She possessed neither paper nor stamps, and couldn't write English. She preferred not to press for compensation.

At the conclusion of the Cold War, when Kerr McGee vacated their Shiprock plant, they piled huge mounds of uranium tailings 60 ft from the San Juan River, the only water source for 15,000 people. When the United Nuclear Corporation mill tailings dam burst in Church Rock, N. Mex., releasing 93 million gallons of radioactive water and 1,000 tons contaminated sediment into the Rio Puerco, the largest industrial release of radioactive materials in continental U.S., the owners refused to provide food or water to the affected population, forcing them to consume contaminated food and water.

Because of their heightened incidence of cancers, lung disease and birth defects, the Diné eventually became activists. Through their efforts, the Diné Natural Resources Protection Act of 2005 was passed, prohibiting uranium mining on lands belonging to the Navajo Nation.

Inupiat
Dumping, Point Hope, Alaska.

Project Chariot, one of Edward Teller's brilliant ideas, was designed to use nuclear explosions to make a harbor for ocean going vessels at Cape Thompson, 31 miles south of Point Hope. 15,000 pounds of radioactive soil was dumped there after the project was abandoned. Whereas before the dump, the death rate from cancer was 4 in 1,000, between 1969 and 1993 it escalated to 57 per 1,000 per year. Inupiat attempts to discover the cause were met with outright lies by the Army Corps of Engineers and other investigating officials, denial by Alaska Senator Stevens, and with stonewalling by the white-run Native Health Service. Not until 1995, and after the Inupiats' vigorous attempts to obtain redress from Washington, did the government medical authorities conclude that the high rate of cancer was caused by local radioactive pollution.

Moslem Bashkir and Tatars
Chelyabinsk 40, Chelyabinsk 65

Both closed cities, both left off official USSR maps until 1989. The Chelyabinsk weapons facility known as Mayak Chemical Combine, is located in Chelyabinsk 40 (now renamed Ozersk). It is a sister-plant to the Hanford, Washington weapons grade plutonium weapons plant. Whereas between 1944 and 1972 Hanford released some 25 million curies of radioactive contaminants into the atmosphere, Chelyabinsk released at least 120 million curies, contaminating the nearby Techa River used by the string of village settlements located along its banks. Whereas Hanford contaminated lands used for hunting, fishing, and gathering by the Nez Perce, Umatilla and Yakama tribes, one of the USSR villages, Muslyumovo, was populated by a majority of Moslem Bashkir and Tatars. In the early 50s, after Mayak had dumped plutonium into the Techa, doctors observed a marked increase in diseases and death in the villages along the river. Twenty-two were evacuated, but even though it was located in the same string of villages, Muslyumovo was not.

In 1957, after its cooling system failed, one of Mayak's 80,000-gallon storage tanks exploded, releasing 20 million curies of debris over some 6,000 square miles, affecting a population of 270,000 people. More than half were evacuated only eight months afterwards, meaning during all that time they consumed radioactive food, including the 1957 harvest. People described how they had to slaughter their animals the night before they left, part with their family heirlooms, and prepare their homes for burning. According to one researcher, the current dose of radiation absorbed by Muslyumovo residents is ten times higher than acceptable levels. 82 percent of the children suffer from memory loss, attention deficit disorder and exhaustion.

Although there is evidence that the CIA already knew about it in 1976 and perhaps before (Bertell. *No Immediate Danger,* 175), the 1957 Khyshtym explosion (named after the nearest town) was never brought to light until 1989 (three years after the Chernobyl disaster) when the local populations finally realized the consequences of living in close proximity to a nuclear weapons facility. Until then, they had referred to their ailments as the river disease. From 1950, incidents of leukemia along the Techa increased by 41 percent. From 1980 to 1990, all cancers rose by 21 percent, and diseases of the circulation system by 31 percent, but although dramatic, these figures remain unreliable because doctors were pressured by the Soviet government to observe quotas when issuing death certificates, effectively capping diagnoses of cancer and other radiation-related illnesses. Despite the advent of the widely-applicable Chernobyl law, mandating the relocation of Chernobyl's victims, and providing monetary compensation, the 444 Muslyumovo families who agreed to move to safer ground still await funding.

Mormon Ranchers, Downwinders, Nevada Test Site

From 1951 to 1962, a population of some 100,000 largely self-sufficient Mormon ranchers tended their sheep in Utah, Arizona and Nevada, downwind from the site where more nuclear weapons were detonated than any other place on earth. "We had cows...that we milked, we drank all the milk...made butter and cheese. We didn't have indoor plumbing...[or] a telephone. After a bomb there would be the fallout, fine like flour, kind of grayish-white. We would play like that was our snow" (*Half-Lives & Half-Truths*, 194).

Not long after testing started, people living downwind began to suffer a long list of health problems. Thousands of sheep became sick and died. One rancher put it like this: "Have you ever seen a young animal that was born completely rotten? I did....Have you ever seen a young animal that...didn't have one speck of wool on their bodies? They were transparent and you could see their little heart a-beatin' until they died. I never seen anything like that...from the time I was a little boy..." (*ibid.*, 198).

The radiation that killed the sheep caused a spike in birth defects. One father remembered, "When my boy was born...his face was a massive hole and they had to put all these pieces of his face back together. I could see down his throat, everything was just turned inside out and it was horrible. I wanted to die. I wanted him to die. I didn't want him to live because I thought there was no way that he could ever make it. I remember going outside the hospital, laying on the grass and just crying and sobbing over it" (*ibid.*, 198).

Marshall Islanders
RMI and Yanomami, Brazilian Amazon

Medical Experiments. When the AEC Advisory Committee of biology and health met in 1956, Merrill Eisenbud, speaking of the Marshall Islanders, made this statement: "...while it is true that these people do not live...the way Westerners do [as] civilized people, it is nevertheless also true that these people are more like us than mice," his remarks presaged how the U.S. would use the Marshall Islanders in an experimental role similar to that of laboratory animals (*ibid.*, 25).

AEC-funded research in the Marshall Islands tracked the movement of radioactive elements through the food chain and the human body, and recorded biophysical changes relative to exposure to radiation. These studies were correlated with studies done in the Arctic, Andes, and notably with biomedical anthropologist Neel's research among the Yanomami in the Brazilian Amazon where, among other practices, he injected his subjects with radioactive isotopes.

Neel reappeared in the Marshall Islands where between 1946 and 1958 the U.S. Government carried out 67 atmospheric atomic and thermonuclear tests, conducting an AEC-funded population study to determine the genetic effects of radiation. A further AEC-funded initiative to collect urine and blood samples was designed to determine the carcinogenic action of ingested and inhaled radioactive material.

On March 1, 1954, the people of the Marshall Island atolls were exposed to radioactive fallout from a 15-megaton thermonuclear weapon. Evacuated 3 days later from Rongelap, the people were enrolled in long-term radiological studies. They were among the first Marshallese to be included in American tests on human beings exposed to excessive doses of fallout in which significant burns, hair loss, depressed blood cell and leukocyte counts and flu-like symptoms, nausea, and radioactive urine were documented.

Three years later this population was returned to Rongelap, to their highly contaminated environment, where they were subjected every year to medical experimentation, unaware until 1982 of the danger of living in such an environment. Additionally they were exposed to biological weapons testing in 1968. In all, the Rongelap population hosted 73 biomedical research visits involving post-mortem harvesting, and in addition to sample collection, very painful procedures such as tooth extractions and bone marrow sampling.

I am indebted to Barbara Rose Johnston and the authors of *Half-Lives & Half-Truths,* for these summaries and for this abbreviated quotation from a letter by a Rongelap subject to one of the AEC-funded experimental doctors: "...your entire career is based on our illness....You have never really cared about us as a people—only as a group of guinea pigs for your government's bomb research....For... the people of Rongelap, it is life which matters most. For you it is facts and figures....We don't need you and your technical machinery....We do not want to see you again....We want medical care from doctors who care about us, not about collecting information for the U.S. Government's war makers" (*Half-Lives & Half-Truths,* 194).

Ultimately, although the people of Rongelap requested that the Nuclear Claims Tribunal recognize the human subject experimentation as a harm, the Tribunal has yet to issue findings or judgments in their case.

With the advent of the Cold War concept of mutually assured destruction, it becomes clear that nuclear weapons and their use directly conflict with the practice of democracy. With its "arrogance, ignorance, limited vision, secrecy, lack of accountability, and lack of compassion…," already with the first use of thermonuclear bombs, the nuclear industry began to show its true face (*Half-Lives & Half-Truths: confronting the radioactive legacies of the Cold War,* 300).

The perverse thought occurs to me: in the Soviet Union, the "Cold War," may have indirectly resulted in as many as one million deaths; and the United States and its protectorates may have suffered just as many. But in this "cold" kind of war, these deaths were not inflicted by the "other side;" they were inflicted on people by their own governments, governments who borrowed from races of color, and the otherwise marginalized, in a kind of tokenism using live subjects instead of the abstract chess board of ritualized war, except in the case of this auto-war, these were real people.

Because both the Japanese government and the U.S. Government have mandated an information blackout on the degree of radiation being released from Fukushima, we now share the condition of the Inupiat of Point Hope, Alaska, the Yanomami of the Amazon, the Diné of Nevada, the African Americans of South Carolina, the Marshall Islanders of Rongelap, and many others: we are being exposed without any say, and we don't know to what degree, nor do we understand the future health threat to us in terms of heightened incidence of cancer and birth defects brought about by chromosomal damage.

We are all Fukushima now.

81.

Our Meat, Our Vegetables, Our Fish, Our Rice…

One hundred and ninety-three days following our planetary catastrophe, while politicians mouth their criminal platitudes, while the skies become saturated with unprecedented moisture, while the rich rob the poor with unrelenting purpose, in continental United States we begin to hear the faint echo of Saori Matsuoka's words, words we first heard before our arrests at Lawrence Livermore National Laboratory on Hiroshima Day. "Our meat, our vegetables, our fish, our rice—everything we eat is contaminated. Vegetables, fishes, meats, and so many other products contain too much toxic to sell or eat. Even tea picked in Shizuoka, where is more than 200 miles away from the power plant, are restricted in the De Gaulle air port, in France, for containing excess toxic."

On Day 193, a report trickles out: peaches purchased somewhere at a local Santa Monica grocery are highly contaminated. (Of course, we can still shop safely at any corporate-owned supermarket.) Otherwise we have a total media blackout. And despite our tax dollars at work, since May, at the request of the Japanese government (among others), the EPA stopped reporting any radnet readings, but it is powerless to stop the trade winds, powerless to stop the swirling plumes of radiation worse than the fallout from 28 atomic bombs. All the Northern Hemisphere is blanketed. When authorities speculate Fukushima will spew its scourge for the next ten years, perhaps they err on the optimistic side. Our soils, our water, our air is being contaminated. Nothing we eat, or drink, or breath is from the previous world, the world before Fukushima. TEPCO says it will have shut down the reactors by 2012. Add ten more years to clean up the plutonium. Removing the melted nuclear fuel from the crippled reactors at Fukushima Unit 1 should start within 10 years after "cold shutdown" has been achieved, and decommissioning may take 30 years or more.

Closer to home, California's two nuclear power plants, San Onofre, within 50 miles of both Los Angeles and San Diego, and Diablo Canyon in San Luis Obispo, both sit on major earthquake faults. Efforts to shut them down have been on-going for some time, although *The Los Angeles Times* failed to cover the recent California

Energy Commission hearings of July 26, 2011, in which the USGS testified it was no longer certain that the California reactors were capable of withstanding future major quakes. If San Onofre or Diablo Canyon were to experience a nuclear catastrophe, it would contaminate California's food supply. "Grown in California," would become a label to be shunned the world over.

I know this. I carry it with me wherever I go. I see the same things those around me see, but with changed eyes. The days now are sunny once more. The dense fog that damped our morning skies has vanished. We are now approaching the fall equinox. At noon the heat is barely tolerable. The breeze is gentle to the skin. Mothers rushing home for lunch, push strollers with nodding infants asleep sitting up. At the supermarket, shopping carts clang and rattle as people return them to their stations. On the main thoroughfares, traffic whooshes by, sucking turbulent air in its wake. Gardeners stir up dust and leaves with gas-powered blowers. Millions of hot particles swirl in the air.

In Japan, children wear surgical masks, ineffective against hot particles. Here, the HEPA mask manufacturers are not yet powerful enough to pressure the press to report on the non-reporting of the EPA. No one seems to want to ask the question. We want to cling to our blue skies a little longer, to our outings to the beach, to the rock-strewn coastlines whose majesty enthralls our eyes, to the redwood forests, our gift from before history. Let us see these skies, these waves, the splendor of these trees a little longer. Let us lull ourselves a little longer till we wake.

In the streets, people run, their ear buds insulating them from other sounds; pregnant women vaunt opulent bellies under the flimsy clothing of an Indian summer. Homeless people extend their cups. An old man sprawls, asleep on the sidewalk, his cane at his side. Saturday people swarm like flies to the shopping malls. A clerk urges a discount coupon on a customer. "Take it. There's only one more hour left to shop, one hour left before the sale ends." In the public park, pensioners toss popcorn to the pigeons.

And on the same Day 193, the Occupy Wall Street newly born movement speaks its first words: "We are the 99%," and "mic check." At Bank of America, OccupyWallStreet protesters camp out, ready for the cop's baton, the tear gas that comes next.

For eyes that see, a proscenium has opened. Behind the curtain, there is no stage, just the pageant of the joggers, the women rushing home to lunch. I watch them breath the poisoned air, drink irradiated water from their eco-friendly water bottles, share a peach perhaps. What were once life and the signs of life, have become Endgame. Our water is poisoned, the water we once drank to refresh ourselves. The soils are poisoned, soils that once produced our food. Our air is poisoned, air we breathed without a thought. But we can't stop. We are made to breath. We are made to eat, to drink, to slake our thirst. And we are made to poison. "In a moment all shall be changed." In Greek, Paul's words read: en atomos. And although we are all Fukushima now, the dancers don't yet know it. They don't want to know it. They are not yet ready to know it. They don't want to learn the steps. It's not true yet. They still have hope. And change you can believe in. They are still children of springtime, and even though it's early fall, they're still treading the rhythm of a gentler minuet. It's what Joanna Macy calls the "double life" (Joanna Macy. *Despair and Personal Power in the Nuclear Age*).

82.

Hope You Can Believe In…

And now, ladies and gentlemen, get ready to place your bets! Extremophiles are lined up at the starting gate, those cute little one-celled organisms that can live through anything, almost anything at all, including a homogeneously irradiated planet! Count all the super-fillies waiting for the next four-billion-year handicap to open:

alkaliphiles – like it alkaline
acidophiles – like it acid
cryptoendoliths – like narrow, microscopic spaces
halophiles – get off in salt ponds
hyperthermophiles - like it hot
hypoliths - like freezing deserts best
lithoautotrophs – get off on mineral compounds
metallotrophs - lap up metals

oligotrophs - feed on starvation diets
osmophiles - like it sweet
piezophiles - want it heavy
cryophiles - like it shivery
xerophiles - like it dry

This time around the smart money's on the radiotrophs, a class of extremophiles that gobble up ionizing radiation, both ultraviolet and nuclear. Right now they're not much to brag about: no nucleus, nothing yet in the way of organelles, just a cell membrane like any other run-of-the-mill archaea. But give these babies a billion years or so, and they just may win the race. There's no telling which appendage may come first: a stalk eye, a webbed claw—but whatever it is, they won't be sliding on their hairy skin by then! And when they're fully grown, I'll bet their moms will think they're pink and plump and perfect.

83.
That 38-Hour Margin of Error Keeps Us All Safe

On the 205th day following our planetary catastrophe, TEPCO proudly announces that from now on, should there be any further interruptions to its cooling system, it expects backup systems to resume water injection within the first 3 hours. But the failure margin is a generous 38 hours after which fuel rods could start melting, releasing a second wave of radioactive fallout over the entire planet. The report makes no mention of the structural defects affecting GE's Mark I boiling water reactors.

Their happy announcement notwithstanding, TEPCO now admits that its emergency systems manual, drafted on the assumption that the emergency generators, including diesel backups, would keep the cooling systems going no matter what, turned out to be completely useless. It says nothing about who exactly authored the manual.

Meanwhile, in Iwaki, 50 km south of Fukushima, the Hawaiian style spa, resort and theme park have reopened for business. Free bus rides are being offered, direct from Tokyo. Seventy percent of the rooms are already booked, and the 28 prize-winning hula dancers who, along with Lady Gaga, won the Japan Tourism Agency's recent award, will present their Polynesian show to a full house.

84.
Concrete Evidence

Picture it: concrete—to the far horizon where the earth etches its parabola against the hectic glow of a setting sun. Even from an elevated place, even seen from an airplane: concrete as far as the eye can see, this, home to some 35 million people; gray thoroughfares, gray high-rises, towers, high frequency antennae, the city spills its gray pall against the day's last light: Tokyo, tombstone of the world.

Once called Edo, where perhaps as few as 100,000 people made a city, crowded with bamboo, glazed roof tiles, gardens, theaters, brothels, temples, pagodas, wooden, arched bridges; where sweepers, and cooks and vegetable boys bustled, their yokes balanced on their shoulders, carrying splashing tubs of tofu, seaweed, fruits of the sea; where in the theaters, to the wail of the shakuhachi, fantastically robed actors, eyebrows shaved, faces whited out with rice powder, re-enacted a drama of demons and ghosts. Now thirty-five million people huddle here, in a wasteland of concrete.

Small images flash on the monitor as I watch, scenes of tranquil houses, sunlit streets, children playing ball, a masked woman stoops in her garden, collecting samples, innocent backdrops for two men who sit on a park bench. One shows the other a map. His finger points to small dots, red, and yellow, and green, innocuous as a child's pegboard game. In a shuttered house, a mother offers a banana to a child, her mother's eyes lit with love for this, her child, her child's mouth filled with pulpy stuff, her eyes filled with trust for the woman who loves her. Mother whose eyes tell the story: How her child sleeps as though she will not wake. Ever. The dark circles under her child's eyes. The first signs of radiation sickness.

This is Tokyo, capital of a small island nation. Its 35 million people may have to be evacuated. But where? The innocuous dots, red for the hot spots, yellow for the spots that are not so hot, but still not habitable, green for the places that once were habitable zones. They are saying, these two men, sitting on a bench, in a playground, in a vast city of concrete, they are saying Tokyo may have to be evacuated.

The small girl reappears, her mouth full, her eyes with the dark circles under them, gazing in her mother's eyes, lit with trust for the woman who withholds a terrible truth behind her eyes.

This is how we will remember the world. The mother's gaze, the child's look of trust. This is the moment when the heart of the world stops. This day, the 206th day following our planetary catastrophe, the day environmentalists chose to designate World Habitat Day.

There was no place more wonderful than this. There was no place more marvelous.

85.

Rems for Rads and Parts For Particles: Homage to George Carlin

> *Justification for safety enhancements at nuclear facilities, e.g., a compulsory backfit to nuclear power plants, requires a value-impact analysis of the increase in overall public protection versus the cost of implementation. It has been customary to assess the benefits in terms of radiation dose to the public averted by the introduction of the safety enhancement. Comparison of such benefits with the costs of the enhancement then requires an estimate of the monetary value of averted dose (dollars/person rem).* (V. Mubayi, et al. "Cost-Benefit Considerations in Regulatory Analysis.")

Did you ever think how much your cancer weighs? I mean, everyone who has a cancer gets it taken out. You either get the surgeon to do it, or if things don't go too well, the mortician gets to handle it. Did you ever wonder where they put the things they take out of you? Let's say you're LUCKY. You only have a SMALL tumor; first it was a ping pong ball, then maybe it grew into a tennis ball. They take it out.

But now the NRC needs to know how much it WEIGHS. Pretty twisted. But it's YOUR TAX DOLLARS AT WORK. OK. You think they weigh it, put it in a jar? The government wouldn't pay for the jar or the formaldehyde. That would taint the statistical results. They just want to know how much it WEIGHS. Now why would the NRC want to know that? Because they're ADDING YOUR CANCERS UP, that's why. It's not enough they get your MONEY, now they want your CANCERS, too. FREE. They're not paying you for any of this. When you lie on that table, you relinquish all your rights to your personal cancer. It's part of the P.A.T.R.I.O.T. Act. So when the government gets enough cancers, and the government weighs them, they get a base line. That's how they can tell how many cancers it takes to pay for a NUCLEAR ACCIDENT.

See, you've got this run down plant. Maybe it was supposed to go off line after 40 years, but it still looks pretty good. Steam's coming out of it in a nice white plume every time they vent it; it photographs pretty well against the fault line. Maybe add another ten. And maybe ten after that. But like you, it's got DEFERRED MAINTENANCE. You got deferred maintenance—you need a new hip, or a hearing aid, or a stent in the old pump. It needs maintenance, too. The hearing aid costs you maybe four thousand dollars. You know how much it costs a nuclear plant? MILLIONS and BILLIONS OF DOLLARS just to ream out that clogged cooling system. But they're not going to pay that! They're the OWNERS! It's too much MONEY.

So they call their boy in Washington. Bring the guys at NRC a nice case of scotch. Johnny Walker. Yeah. Write a check. It's pennies on the million. Chalk it up to overhead. They get a law passed. And the law says they don't have to make those repairs if they COST MORE THAN ALL THE CANCERS THEIR LITTLE ITTY BITTY NUCLEAR ACCIDENT WOULD CAUSE.

So how many cancers is that? You think the NRC is gonna weigh all those cancers to assign a dollar value? Nah. They just go to the Department of Health, get the bean counters to make PROJECTIONS. Take Three Mile Island. Ten years down the line, maybe 550 cancers, 20 years down the line maybe 1100 cancers. Any kind of cancer. So then they add them up. The guy in Washington sends around some champagne, some Iranian caviar (hate Ahmadinejad, but love that caviar!). Crunch, crunch, crunch goes the Department of Health. Crunch, crunch, crunch goes the NRC. Then they have a joint session at the RESORT: girls! Champagne, MORE CAVIAR. Crunch, crunch, crunch. The figures come out. Taking care of the cancers only comes to ONE MILLION THREE HUNDRED EIGHTY-THREE THOUSAND DOLLARS AND THIRTY-NINE CENTS. But their projected repairs come to TWO BILLION, ONE HUNDRED FORTY-THREE MILLION AND FORTY-THREE CENTS. So everybody's happy: the One Percent gets to keep their rusty pipes. The doctor gets his cut. Us 99 Percent get our cancer taken out. And if we're still alive—and we still have a medical plan—we get a jar free to take it home.

86.

Waxy White House Buildup

The NRC schedules a public hearing for Friday, the 210th day following our planetary catastrophe. The Nuclear Information and Resource Service sends out an announcement: the NRC is soliciting comments from the public—there were so many phone calls last time, they crashed the phone lines—they had to reschedule another hearing for Friday. NIRS internet announcement makes it easy: if you can't be in D.C. in person, maybe you'd like to urge the White House to retire every one of the 23 defective GE Mark I reactors from active use in continental USA. Dale Bridenbaugh, the design engineer who quit GE many years ago because he knew they were defective, is going to testify.

Since your government became dysfunctional, you've signed internet petition after petition. You're so retro, you imagined your small CLICKONSEND might make a difference. If nothing else, it's demonstrated how your government's gone fishing—dialing for dollars—with no one left to mind the store. But this one, this last petition—you're going to sign it anyway. The horse may have gotten out of the barn—you suspect as much; there's enough radiation swirling around the Northern Hemisphere to guarantee bee colony collapse—but at least you can sign that petition to the White House even though NIRS warns, "one person reported that it took her 39 tries, but the system seems to be working better now."

Computers give you PTSD (you were born when people just started having radios in their living rooms), but you hallucinate for a brief moment that this White House really wants to hear from you. You click on the link NIRS provides. Up comes the screen. But before you get to sign, you need to register. You click on the link. Three bands appear: "name," "first," "last." Directly in front of them, is a band aid. You don't have to be told your government has a booboo, but there's no way you can push it aside because it's a virtual bandaid. You can't drag it either. Result: you can't register. You can't sign the petition. Fooled you once. Try again. From zero. Click on the link. No bandaid! You sign your name, first and last. You post your e-address. Now it says, "put in your password." Password? OK. You put in your password. Click.

The dread pink window appears: you have entered a user ID that is invalid. It goes on to tell you how badly you goofed, how you'll never get it right. Your password is wrong. Your name is wrong. Once you walked out on a guy over something like this.

Back to the drawing board. *Nuh, unh!* Now you need to get a new password. You click on the button. *Not yet, not yet.* They're going to e-mail you a new password. You sign off. You wait. Here it comes! The bright and shining e-mail from the White House. Billclintongeorgebushbarrackobama has sent you—yes, you!—a personal password! An e-mail from the Leaders of the Free World. The most powerful Entity on Earth. To you! You click. There it is: Y23Bxdfu379K. That's all you have to remember. That's your new identity. It's what allows you to vote in this public opinion poll. The White House will surely take it to heart when it hears from Y23Bxdfu379K. You get to sign the Petition. You read the list of all the other Y23Bxdfu379Ks who probably spent 25 minutes like you have trying to communicate with the White House and the NRC: *the defective GE Mark I BWR must be Banished from the Earth*—although it's already done its lethal damage last March 11th.

You even get to see the last Y23Bxdfu379K to sign is the 3,445th signatory. You've already spent 25 minutes doing the work your government should be doing. You wonder does Billclinton-georgebushbarrakobama spend 25 minutes on every decision that comes to hishishis desk? When your government's out to lunch, no one has time to be president any more.

Three thousand four hundred and forty-five people plus you makes 3,446. And say that each, like you, spent 25 minutes jumping though the Earwax Hoops. Do the math. That's 86,150 person-minutes. Or 1,439 person-hours, or 60 person-days—two whole person-months! Two person-months of human effort to try to turn the Ship of State that's out to lunch around! You know there's only one Ship of State. But to turn this Goliath around, you need 99 tugs. And there's all that waxy White House build-up to make sure it won't happen.

87.
"This is a Matter of Life and Death"

On this, the 210th day following our planetary disaster, I have been up before dawn calling one of the dedicated conference lines where I listen in—just in time to hear the impassioned testimony of Michael Mariotte, Executive Director of Nuclear Information and Resource Services who's testifying before the Nuclear Regulatory Commission hearing. (The band aid he's referring to below is that troublesome torus vent, designed to release radioactive high-pressure steam generated in a severe accident by allowing the unfiltered release directly into the atmosphere.)

> *In 1986, we determined that the GE Mark I would explode with a 90% failure rate. You chose a band aid approach, allowing the Mark I to vent small amounts of hydrogen. Rather than solve the problem, you put a band aid on it, trusting to a defense of dumb luck.*
>
> *That dumb luck ran out at Fukushima, where the reactors exploded as predicted 40 years earlier, and where eighty thousand [families] in Japan lost their homes. Now you're proposing to make the venting system even better, by adding*

another band aid on top of the first band aid, crossing your fingers for continued dumb luck. Why don't you just close them? If they were closed, no one would notice. They don't supply that much energy. Instead you want to continue relying on band aids and dumb luck. We are tired of that kind of thinking—putting profits above environmental safety, and the safety of our homes, and our families and our children.

That's why ever-growing numbers of people are Occupying Wall Street, and why that movement is spreading to city after city. We're fed up with the concept that the interests of large corporations consistently are placed above those of the American people. This is a matter of life and death. It will affect our children and our grandchildren.

No one knows what the next challenge may be to the Mark I BWR, but we know what will happen. The reactor will fail. You have the power to close down those reactors. If you don't shut them down, we may just have to go in there and shut them down ourselves.

One more speaker, someone from Georgia WAND, is allowed to speak before the lines go dead. I can hear the voices of three more speakers still waiting their turn. One of them, a woman, comments: "At the last hearing, when one of the commissioners asked what to do with the remaining lineup of speakers, one of the other commissioners said: 'Don't answer them.' Clearly they don't want to hear from us."

If you go to the website of the NRC, you can read:

The NRC Approach to Open Government: As an independent regulatory agency that prides itself on openness, the U.S. NRC is pleased to take an active role in President Barrack Obama's Open Government Initiative EXIT *with its focus on open, accountable, and accessible government. The NRC has a long history of and commitment to transparency, participation and collaboration in our regulatory activities.*

We should have hit the EXITs long ago, perhaps from the beginning.

88.
Little Essay On Elegance

I am not a scientist; I have no mathematical pretensions, but in mechanical design I know the most elegant designs are engineered to guarantee the most efficiency—the least loss of energy, or heat. I know enough to understand that design kinks need to be worked out before the foundation is laid, and that they can't be addressed on a Pincus-the-Tailor basis as the current system of nuclear plant licensing actually allows.

I refer to Pincus, the Tailor, joke #418:

> *Yossi goes to Pincus, the tailor, to try on a new suit. The first thing he notices is that the arms are too long. "No problem," shrugs Pincus. "Just bend them at the elbow and hold them out in front of you. See, now it's fine." "But the collar is up around my ears!" "It's nothing. Just hunch your back up a little...no, a little more...that's it." "But I'm stepping on my cuffs!" Yossi complains. "Nu, bend your knees a little. To take up the slack. Look in the mirror. What did I tell you: Now! that's a smart suit!" So, twisted like a pretzel, Yossi makes it to the street. Gimpel sees him.*
>
> *"Oy," says Gimpel, "Who made you that suit?"*
>
> *"Pincus, the tailor."*
>
> *"I knew it! Only Pincus could make a suit for a cripple like you."*

In my own life, absent mechanical training, I tend to look for elegance in nature. In asclepias fruticosa, for example, whose fruit, a crazy, spiked, chartreuse balloon worthy of Dr. Seuss, houses a Fibonacci distribution of brown-robed seeds, each in its monkish cell of fibrous white, in parallel monastic corridors where each one is equally exposed to the sun's filtered rays.

I find examples of true design elegance useful, two in particular:

- MIT has come up with a design combining nanotechnology, solar power and robotics in a single device designed to scoop oil spills from the ocean surface. Make no mistake, I am not advocating oil spills, let alone a fossil-fuel economy, but MIT's technological breakthrough is attention-worthy (Alyssa Dangelis. "Solar-Powered Robot Swarm Could Clean Oil").

•Michael Pearce is a Zimbabwe-based architect whose building designs incorporate biomimicry of creatures—in this case termites—that have solved technological challenges millions of years ago. His Eastgate Office Complex generates 90% of the power it needs in its building systems (Inhabitat. "Green Building in Zimbabwe Modeled on Termite Mounds").

Once you see what good design looks like, you tend to dismiss everything lesser as schlock. That includes the GE Mark I BWR, the model that failed at Fukushima. The United States has 23 GE Mark I's in operation, and none of them have been decommissioned—yet.

89.

Shut 'Em Down

Protesters from the UK, their ranks swelled by people from downwind countries like Belgium and Germany, recently came together to block access to Hinkley Point, one of eight sites designated by the UK's Department of Energy and Climate Change (DECC, double counterpart of the U.S. NRC) as suitable for new reactors.

Although STOP NEW NUCLEAR is a coalition built on the shoulders of several UK anti-nuclear groups, it knows what democracy looks like, and it knows how to block a road. There's room for fun and art in this movement. Here they come, a colorful stream of folk led by a theater troupe, horns blaring, drums pulsing, their colorful gonfalons held high. Suddenly sirens blare, a factory whistle blows off steam, and what first appeared to be a colorful band of folks hits the ground, collapsing in death throes, re-enacting the horror of Fukushima's nuclear disaster.

They are stopping access to the power station where EDF plans to build a new reactor, swelling the wave of the global revolution that sparked first in Iraq in protest of NATO atrocities, and in Tunisia where Mohamed Bouazizi immolated himself when police confiscated his vegetable cart and his weighing scales. They are the do-it-yourself shut-it-downers, part of the 99% popular revolt. "This is the start of a new movement," proclaims Andreas Speck, dusting himself off, "a celebration of resistance against the government's and Electricité de France's (EDF) plans to spearhead the construction of eight new nuclear plants around the UK."

In the US, the NRC chair, Gregory Jaczko, claims that "events like Fukushima are too rare to require immediate changes, that stuff like that hardly ever happens." Should you have any lingering doubts, you may care to consult the mind-numbing April 13 "Expanded NRC Question and Answers related to the March 11, 2011 Japanese Earthquake and Tsunami," a 50-plus page conveyor belt of PR bromides and infantilizing double-speak (Gregg Levine. "NRC Chair Jaczko: Event Like Fukushima Too Rare to Require Immediate Changes").

Commenting about the recent string of accidents in U.S. nuclear plants, John Lane, NRC's senior reliability and risk engineer, compared the situation to a flip of the coin. "While you would expect to get a head half the time if you made dozens of tries, it's possible there could be a surge in heads over a shorter span of tosses. If you flip a coin ten times, you're liable to get 6 or 7 or 8 heads, and our feeling is that's essentially what it is...." Lane's response, rich in gravitas, seems pretty emblematic of the NRC's approach, but undoubtedly the 200,000 Japanese people who have lost their homes as a result of the failure of GE's defective Mark I reactor might view a coin toss rather differently.

Not long after I'd graduated from reading *Winnie-the-Pooh*, I discovered that a long line of everybody is hardly an expotition. In the French films of the late 40s and early 50s, you could see long lines of refugees clogging the roads, their possessions tied in bundles, or piled high on the beetle motorcars of the 30s and 40s, fleeing Nazi Messerschmitts coming in low to strafe with deadly accuracy. In the ending frames of Jules Dassin's *The Greek Passion,* Greek villagers take positions in the alleys, between the narrow apertures of houses, rifles pointed at the long line of refugees fleeing the Turks. And perhaps most dire of all, Satiajit Ray's film, *Distant Thunder*, depicts the lines of famished people that never stops streaming over the hillsides, fleeing the starvation of the Bengali famine of 1943. A last, in 2005, real life brought the line to my Oakland doorstep with the black folk fleeing the worst hurricane ever to hit the shores of the United States. These were folks whose hands I held in prayer, whose stories I listened to, whose tears I helped stem, who came by the hundreds to the Bay Area from the nightmare of the New Orleans Superdome, the New Orleans Convention Center, the Houston Astrodome, traumatized by the waters that never seemed to crest, by the endless limbo of sleeping on soggy cardboard, huddling by the thousands under bridges, dying of thirst on the overpasses. These same folks are now being dunned by FEMA, which demands they pay back the few miserable dollars in emergency funds FEMA granted them six long years ago.

I know firsthand about long lines of everybody. It doesn't matter that an ethnic group may have been acculturated through centuries to practice emotional restraint; deracination scars all equally.

I receive an e-mail from my Japanese pen pal:

> This real situation is just beyond human scale, as you said, and we can't have any prospect about the consequence of the devastation of Fukushima Daiich nuclear power plant.
>
> But please understand our thoughts for our hometowns. About 200,000 people in Fukushima might have to leave their homes. Seeing objectively, it is so much difficult for them to return to their hometowns for a very, very long time. Perhaps, almost all of old people will not be able to return for their living in this real world. But, they never give up the return, even though it becomes far much longer after all old people died. We have no land to cast away in our country.
>
> In friendship,
> Narumi

If there's to be a change in the world's nuclear climate, it will not come from governments (with the possible exception of Germany's), certainly not here in the U.S., where our Secretary of Energy, Stephen Chu, minimizes the impacts of radioactivity, as do many atomic physicists associated with the national laboratory system, and two of Obama's top White House aides, Rahm Emmanuel and David Axelrod, were instrumental in bringing about the corporate merger that resulted in Exelon, and they've been writing PR (more bromides) for Exelon, a corporation which has come to be known as "The President's Utility."

Instead, change will come from the 99% and from activists and local people in a life-and-death 40-year-long fight to keep the earth from becoming a designated exclusion zone—activists like Shaun Burnie, who in a recent video describes Greenpeace's initiative to block a shipment of weapons-grade plutonium fuel (MOX) from France and the UK from being off-loaded into the Fukushima reactors in 1999. Video Clips (available on YouTube) show activists, and local people, at great risk to themselves, blocking the delivery ship. Their successful action was followed by a court case—which they lost—and by ten more years of opposition. Although TEPCO loaded MOX into Reactor 3 shortly before March 11, but for the 1999 resistance of Greenpeace aided by local activists, which prevented the loading of many more tons of MOX, the 2011 explosion there would have been infinitely more catastrophic.

Vermont Yankee, one of the 23 U.S. defective GE Mark I BWRs, is located in Vernon Vermont. Although it's due to be retired in 2012 at the end of its useful life, and although the Governor of Vermont has denied is renewal application, Entergy is suing in federal court to relicense it, despite admitting in its 2010 report that Vermont Yankee has leaked 31,800 picocuries of strontium-90 at ground level. On April 22nd, the 41st day following our planetary catastrophe at Fukushima Daiichi—which happened to be Good Friday—11 women (average age 71) chained the gate shut and stretched caution tape across the Entergy driveway, where spray-painted warnings they applied months ago are still visible. Then they read their statement:

> We are here today to shut down the Vermont Yankee nuclear power plant IMMEDIATELY and WITHOUT DELAY and FOR GOOD. We are appalled at the irresponsible

action of the Nuclear Regulatory Commission in granting Entergy permission to operate this dangerous facility for more than another twenty years. Because the federal government and Entergy will not honor the public good by shutting down Vermont Yankee, we must take this action and shut it down now. NO MORE ACCIDENTS. NO MORE LEAKS. NO MORE LIES. NO MORE TAX SUBSIDIES. NO MORE. ENOUGH. Shut it down now.

State *and* local police wasted no time taking the women to the Vernon police station, citing them for trespass, and booking them for a June appearance in Brattleboro's Windham District Court. By some nice coincidence. the date on the citation, April 22, also happens to be designated Earth Day (*The Nuclear Resister*. "Global Resistance to Nuclear Power on the Rise," June 7, 2011).

(On Day 318, Federal Judge J. Garvan Murtha rules in favor of Entergy Corporation, allowing Vermont Yankee to continue operating past its state-approved license expiration date of March 21. The judge ruled that two of the three bills passed by the Vermont State Legislature, Acts 74 and 160, are preempted by the Atomic Energy Act [Kyle Jarvis. "A Federal Court Gives Vermont Yankee the Go-Ahead"]).

But direct action can and *does* make a difference as we saw in 1981 when the women of Cardiff, some 44 of them, walked 125 miles to Greenham Common to stop USAF/RAF installation of a cruise missile base there. And how they danced over the missile silos, and hung pictures of their loved ones on the fences, blocked the gates, tore down the fences, lived tenting under trees through many a bitter winter, and wouldn't go away until they'd made a statement about the deadly missiles and stopped the countryside—and the gene pool—from being poisoned for all eternity by nuclear contamination. They came in their thousands, hundreds of them risking arrest, until they succeeded in closing down the base and making it extremely inconvenient for the government to try taking over commons and other open spaces ever again.

We are seeing that same spirit of resistance go global now with the 99% movement. As Bob Gorringe has written on this, the 210th day following our planetary catastrophe: "The encampments are the conscience of this country." The 99% movement has brought disparity of wealth into the national conversation. Although it may have been criticized for refusing to say what it wants, our conscience makes no demands, because to do so legitimates a government that has largely become dysfunctional. Our conscience uses a "people's mic," a way of repeating a speaker's words phrase by phrase so that even crowds of thousands can "own" the message, and hear that speaker without use of any amplification system. Our conscience camps in tents, under blue tarps rigged against the fall downpours, the winter snows. And at each evening's General Assembly, our conscience makes consensus-based decisions—in a renaissance of democracy.

90.
Nine Hundred Square Miles of Soil

As of October 10, day 213 following the catastrophe that continues to contaminate our planet, the exclusion zone exhibits cesium contamination reaching down 5 cm (or 2 in). Cleanup will require 29 million cubic meters of contaminated soil to be scraped away from some 930 square miles.

And all the grass and fallen leaves in the forests, and all the dirt and leaves from the gutters, will have to be hauled away, and if you piled all the soil, and all the grass, and all the leaves, and all the dirt, it would fill 23 football parks to contain it. Keep that in mind next time you watch a Raiders game. Twenty-two more games and you pretty much have soil contamination over Fukushima and four neighboring prefectures under control. In the absence of 23 football parks, the problem is where to put it, and so far, the Japanese government has allocated only 220 billion yen (or $2,856,980 dollars) to do the job.

There are no more predatory lions and tigers to keep our numbers trimmed. If human proliferation continues unchecked, homo saps will need 27 planets by 2050. Possibly an extra 28th planet could serve as a repository for all the irradiated soils Gaia will have accumulated from all future Fukushimas—up to that point at least.

How many years does it take to rebuild 2 inches of top soil? The answer ranges between one and two thousand years. That suggests a new unit of soil/time measurement: the kyear/squareinch. Imagine now that TEPCO had followed through with its March decision to abandon Fukushima—and all Japan—and the planet—to its fate, leaving a skeleton crew—for security purposes—of 50 people as it actually proposed. Imagine if PM Kan had not succeeded in persuading Masataka Shimizu, TEPCO CEO, not to cut and run. Another 970 square miles—roughly the area of Tokyo—would have had to be hauled away, another 23 football fields of soil would have needed to be scraped. But where do you put 35 million people—the population of present-day Tokyo? And where do you put the rest of a contaminated planet beyond Tokyo's 970 Square miles?

91.

Meditation on the 273rd Day Following the Planetary Disaster at Fukushima Daiichi

I am trying to understand what it means to inhabit Earth. What does it mean to recognize the gift of Light?

How can it be that this Gift goes unrecognized? How can that be?

Does it depend on where you opened your eyes at the moment of birth, when all things were without name, without need of explanation, unentangled by words?

What were the first things—for you—what were they? Do you remember the light? Do you remember the body that held you, the arms that held you? Do your nostrils remember their smell? When that smell was absent, do you remember how you screamed and wailed and cried? Are those your first things?

Say you had been slow to learn. Would the changing of the light have told you you lived in the shadow of a star? That you were a small mark, a spectator, at the grandeur of its displays? Standing on a small sphere that spun a celestial passage through a wider universe? Would you have spent enough time in one place to discover that you seemed to move to her rhythm, while the star stood still? Would you have come to learn that on your own? In the silence, would the universe have let you hear its hum?

What does it mean to inhabit a landscape in a world that presses forward day by day, bent on doing this and getting that, bent on passing through without paying too high a price: the social contract, the social network, the marketplace, the door closing, pushing your way in the rush-hour throng, buried in the soot and darkness of a subway tunnel going nowhere?

Amidst the trees, in the sands of the red deserts, at land's end where the waves lick at the shore, is this the place the rush-hour stops, where the landscape breathes as I breathe—in and out as I breathe it, where it breathes me?

Whether tree, or desert, or sea, no matter, each landscape needs time, time to live inside it, to discover every grain, every leaf, to learn it as you learned your mother's skin, her smell. At the beginning. Like the First Things that you learned.

Sometimes, in the deep forest, I stop in my path. I close my eyes. I imagine—when I open them—this is the first thing, the first moment I see when I spring new from inside my mother's loins. This splendor.

I am trying to understand being born to the urge to destroy, to rip mountains apart, to pour thousand-year poison into the seas, to belch soot into the sky, to kill everything that lives. Where does it start, this impulse? In what mind? Where is the axle that turns this wheel?

92.
Three-Plus Bags Full

The Japanese mark the 214th day following our planetary catastrophe by receiving a 12-person delegation of IAEA international experts headed by Juan Carlos Lentijo, general director for radiation protection at Spain's nuclear regulatory authority. The team is in Japan to inspect the site of the explosions, and various affected towns such as Ryozen, where some households have been designated as government evacuation advisory sites due to high radiation contamination; the village of Iitate, for example, where demonstration experiments on farmland are being conducted; and the town of Date, where Mayor Shoji Nishida proudly demonstrates that radiation has been reduced down to one-tenth in the school yard after surface soils have been removed.

One member of the delegation inquires where the town disposed of it. The Mayor admits certain problems because there is as yet no formal legal framework addressing disposal procedures. But in the meantime, until the town receives clearer directives, the Mayor explains that the waste is stuffed in bags, and housed—temporarily—at the back of the school gym.

Meantime, 80 farmers in eight prefectures, notably Miyagi and Fukushima, are stuck with 7,200 tons of rice straw, unable to incinerate it because of local opposition. Like the schoolyard dirt, it is being kept on site in barns and nearby storage facilities.

93.
Spreading Trade, Spreading Flotsam, Spreading Trace Amounts of Cesium

The Japan Times reports that for the first time in its history, Japan has occupied a booth at the Guangzhou Trade Fair to exhibit products exclusively from the three prefectures most seriously affected by radioactive fallout: Fukushima, Iwate and Miyagi. There's a manufacturer of lacquerware, another of ceramics, a third of industrial products. Almost as an afterthought, the last paragraph describes a visit by Chinese Premier Wen Jiabao, who checks out some food items and talks with Japanese representatives. Food? From the disaster zone? But there is no further comment.

Another article describes how a Russian tall ship has intercepted an abandoned boat complete with microwave and TV, close to Midway Atoll, 3,000 miles east of Fukushima, exactly where surface current simulations run by the International Pacific Research Center indicate flotsam will pile up. A picture shows the Russian crew throwing a towline to secure the boat, promising to return it to its rightful owner in Japan—if its rightful owner can be found. Crewmembers wear no protective gear. Their handling of the boat is passed off without comment, as though there were no possibility whatsoever that the boat might be radioactive.

IRPC's plotting of the surface currents has been proved correct, but where have the millions of gallons of contaminated water TEPCO released into the Pacific gone? Have they followed the same ocean route as the derelict boat? Since July, seals have been washing up on the shores of Alaska suffering and dying from a disease marked by patchy hair loss, irritated skin around the nose and eyes, and bleeding lesions on their hind flippers (ENENEWS ADMIN. "Diseased Alaska Seals Tested for Radiation").

Closer to home, the Department of Nuclear Engineering at the University of California at Berkeley (http://www.nuc.berkeley.edu/UCBAirSampling) continues air, soils and milk product monitoring. Milk continues to show trace amounts of Iodine, and Cesium-134 and 137, in samples taken as far away as Sacramento. Local soils show contamination as well.

> However, Cesium-137 was detected down to a depth of 6 cm. This Cs-137 is not from Fukushima for two reasons: (1) the cesium from Fukushima came in a one-to-one ratio of Cs-137 to Cs-134, and (2) cesium deposited earlier this year through rain could not yet have diffused down more than a centimeter or two into the soil (cesium diffuses very slowly through soil). As with our other 'older' soil samples, this excess of Cs-137 must be due to previous fallout depositions, primarily from *atmospheric weapons testing* [itals. mine].

This statement refers to a kind of generic weapons testing. Why do I find it strange that it (and many like it) leaves out any mention of agency? Why do I get the feeling that this omission makes such a statement pass as perfectly natural, perfectly matter-of-fact in tone? As if nuclear weapons testing is a foregone conclusion, bought and paid for—in this case by the people whom it happens to irradiate?

Had Fukushima's events not prompted this testing of local soils, we might never have discovered our soil's contamination, or that the nation we once imagined was ours was responsible. Responsible for what? Waging nuclear war on its own soils? On its own people? What people? The people that belonged to it (the "nation")? Or did the nation once belong to the people? Or did it pass from being a British colony to becoming a kind of neocolony for the benefit of oligarchs of the home-grown variety, never to emerge as a nation in its own right?

94.
Bought and Paid For

There's a baffling Old Testament saying: don't boil the kid in its mother's milk. Although there is an interesting (and very convoluted) alternative explanation, most people assume it means if at all possible one ought to avoid acts of such ultimate cruelty as to break the hearts of people who are closely bound together. Full disclosure: it's my personal view that each human being is bound to every other as one flesh—and even further: all living things are bound together. I realize not everyone shares my view.

On this, the 221st day following our global catastrophe, *The Japan Times* publishes a very unsexy article about the 2012 Japanese tax code. According to its tabulations, each family will be taxed proportionally to fund disaster reconstruction. But there is no breakdown in terms of how much the government plans to allocate to repairing tsunami damage, how much is to be allocated for decontamination expenses, and how much for housing and compensation for the 200,000 people displaced from the 12-km exclusion zone. But the unstated point is that Japanese citizens will have to shoulder a burden for which GE—and TEPCO—are in great part responsible. (And as U.S. citizens will have to do in the event of a similar occurrence.)

Case in point: when the Sunfield Nihonmatsu Golf Course sued TEPCO for high levels of cesium contamination, which have made the course inoperable, TEPCO's lawyers, resorting to the arcane legal principle of *res nullius,* argued that "radioactive materials that scattered and fell from the Fukushima Daiichi nuclear plant [now] belong to individual landowners, not TEPCO," which doesn't own them anymore. And although the court threw out TEPCO's argument, the court ruled that cleanup was the responsibility of local, prefectural, and national governments.

If there were to be a similarly devastating accident on the United States to any one of its defective GE Mark I BWRs, such as Vermont Yankee for example, because of the Price Anderson Act, the local citizens would have to pay. It's hard to imagine that GE's culpability would enter into the conversation any more than it does now. In fact, U.S. citizens already pay for their government to listen in on their phone conversations, comb their internet and social media conversations for key words and phrases (hint: don't say or type 'b-mb'); they pay the government to x-ray, and strip them every time they want to fly; they pay for roaming surveillance vehicles capable of seeing inside their homes and places of business; and they pay their municipal police establishments to shoot non-persons at will (here I refer to homeless persons and to persons of color, among others), as they repeatedly do in the San Francisco Bay Area and elsewhere.

They pay for incarcerating 1.3% of their own population—the highest rate of incarceration in the world. They pay for a prison industry in which a full 23,000 prisoners are held in solitary confinement—a practice that the Geneva Convention defines as torture. They pay for 3.1% of their population to be under court supervision at any one time. They pay for their soils to be contaminated, and for their own health to be impacted by practices of the nuclear power and nuclear weapons industry. And they pay the salaries of Supreme Court justices who like to steal elections, and execute at least one innocent man. That's a lot of boiling, a sea of milk and a ton of kids.

95.

And Now for Something You'll Really Enjoy...

Do you ever fantasize about 'sticking it to them?' I mean, really sticking it to them? Here, had I been Lawrence Sterne, I might have provided a blank page where you could fill in your most lurid fantasy of striking back at a sea of wrongs. But to get you thinking, I'll share my stick-it-to-them fantasy or rather my stick-them-to-it fantasy *du jour:* its my invention of a fairly non-descript adhesive suitable (for example) for application along thin blue lines just before riot cops get the command to charge. It stops them in their tracks—a kind of variation on "statues," the kid game we played back when. The glue is also suitable for more personalized applications (for example) on board room seats, and in certain personal hygienic situations.

In the larger (and more high-minded) scheme, I suspect "they" tend to stick it to themselves. For example, do you listen with some amusement to the strident scoldings by Secretaries of State, and UN Ambassadors, pushing democracy on behalf of the greatest terrorist state on earth? No? Perhaps you need to develop a more Eastern European sense of humor. Or presidents waving a stern finger at folks bent on Islamism when they themselves favor multiple targeted assassinations, three of them (by today's count) of their own citizens. That's a bit of a stretch, I admit, but in this particular case, if I had one, I could remove my hearing aid.

Perhaps I set unreasonable hopes on the value of education. I confess to keeping a list of every member of Congress who needs to live in public housing and survive—creatively—on welfare and food stamps for a year. And a list of bankers, whose mistresses would be forbidden for an entire month to shop anywhere but Target, and buy sale items only. No cheating on layaway would be allowed. I have a waiting list of CEOs to be individually chauffeured in luxury tinted-window limousines to the High Sierras, and dropped off in a wilderness location to walk their way out, equipped with three-piece suits, power loafers, and keys to the car. No cell phones or GPS devices allowed. I have a list of generals waiting to be kidnapped dripping wet from their field showers by non-English speaking peasants, armed with pitchforks, machetes, and bushels of jalapenos. (Add handcuffs to make sign language awkward and eating difficult.) And a waiting list of hedge fund managers eager to dine on fresh-minted pasta, with special Sunday specials consisting of penne d'oro laced with American gold eagles. (No gagging allowed.)

But I'm calling on the spirit of George Carlin to stick it to the nuclear industry. It takes a great comedian.

96.
George, Who Are These People...?

When she came home after a long day's work, my mom would see all these people flaked out on her antimacassars and she'd say, "George, Who Are these People? Don't shrug at me. I wanna KNOW. Who ARE THESE PEOPLE?"

"It's OK, mom, they're just your ordinary hoodlums, you know, guttersnipes, your everyday cutpurses."

That was then. Now you go to Washington, what do you see? A daisy chain going from K Street to Congress and back from Congress to K Street, sucking each other off, making the world go round. Now there's a form of alternative energy they ought to try and harness. If it powers Washington, it could power the world. The big utility guys sucking off the NRC guys, the NRC guys sucking off GE, Westinghouse.

So I'm in this watering hole where they have their meeting and I'm watching them sucking up, and guess what?? Their feet don't touch the GROUND! That's right. Their Guccis don't even touch the GROUND. And at the bar these guys don't add water. They suck up everything straight up. They don't eat. Food don't pass their mouth. They don't SHIT! They don't have kids! They don't have parents! They don't even have fucking NAVELS. So I'm looking for the oxygen tanks, you know those itty bitty tanks people wheel along, like they're walking Fido on wheels—no OXYGEN TANKS! So now I know these guys don't have to breathe. BECAUSE THEY DON'T LIVE ON THE SAME PLANET YOU DO. They don't walk on the same ground. They don't eat, they don't SHIT! They just keep on building new nuclear reactors, churning up the cycle: more plants, more bombs;

more bombs, more DU; more DU, more profits. More profits, more plants. More plants, MORE BOMBS.

The only way you stop it is you send these guys to Chairman Mao's re-education camp for aparatchiks. Show them some dirt, some shit. Show them some mud.

"DIRT! MUD! Yuck! What are we supposed to do with that?"

"You walk on it, you grow things in it. You plant the seed. You water it. But you don't watch it grow. Nah. Dig here, plant there. Move right along."

This is how the planet works. And they're all whiney and nose out of joint. "But we were just boiling WATER in those reactors..."

"You gotta be KIDDING! You gotta have a REACTOR, you need MILLIONS and BILLIONS of dollars just to boil WATER? Not here, sonny, you want to boil water in Chairman Mao's re-education camp, you use ALTERNATIVE ENERGY. You don't even get to light a MATCH."

"But where's the profit in that? Where's the MONEY?"

"You don't get no money, you don't get no profit, you just get a planet with dirt on it and worms in it and shit on it that makes things grow. Because, guess what? Otherwise there ain't gonna be no dirt anymore 'cause it was IRRADIATED and they had to bury it in a nuclear waste dump because no dirt was left to bury it in. And there ain't gonna be no boiling water reactors any more 'cause they all blew up. That's why people can't live here any more. Because there ain't no planet anymore. So if you wanna boil water in Chairman Mao's re-education camp for aparatchiks, you set up the FUCKING SOLAR COOKER."

97.

And You Thought Nukes Guaranteed "Clean" Energy...!

If the complete life cycle of nuclear reactors is taken into account, it turns out that nuclear energy actually hastens climate collapse. Although a fully-functioning reactor releases little CO_2, nuclear energy, contrary to its vigorous PR, is a huge source of CO_2 pollution. All the steps required for its implementation require the polluting use of fossil fuels: all the fossil-fueled trucks, drills, locomotives, and cargo ships involved in such processes as mining and extraction of uranium, transportation of uranium to refineries and enrichment facilities, operation of refineries and enrichment facilities; transportation of fuel rods to be installed in reactor cores.

Besides the processing cycle, fossil fuels are consumed in fabricating the plant's thick concrete housings and the huge metal parts involved in plant construction. According to Gar Smith, "it takes many years for a fully operational nuclear plant to generate sufficient energy to offset the energy consumed in the plant's construction" (Gar Smith. *Nuclear Roulette*, 61).

And this is only half the story. Eventually, the operating cycle of these plants will end in one of two ways: either by decommissioning—or catastrophic accident, impacting the entire planet. Either way, abundant use of fossil fuel will be required to decommission or for clean up, and whether slow or rapid, destruction of life on the planet will be guaranteed.

But worse, intensifying climate collapse brings storms, floods, earthquakes, and fires, seriously endangering nuclear plants. In summer, 2011, Ft. Calhoun came within twelve inches of becoming America's Fukushima. Virginia's North Anna came within 1 Richter degree of becoming our Chernobyl.

98.
Without a Navel

> *I realized that nuclear energy, both because of its origin and through its implementation, is rooted in war, oppression, secrecy and manipulation: what is sometimes referred to as 'the rule of the fist.' Because of this, nuclear energy policy [is] a convenience for superpower military strategists, and nuclear energy, whether for war or energy, [is] a single human endeavor.*
>
> (Rosalie Bertell. *No Immediate Danger*, 17)

October 26, the two hundred twenty-ninth day following our planetary catastrophe. At the beginning of life, if one sees at all, it is through a kind of narrow opening, a kind of peering through blinds through which one takes a very limited measure of things. Born in 1932, I was fastened from the beginning to a narrow seam of history—a global economic collapse, the rise of Nazism, the opening of the particular Pandora's box of my turn on the treadmill of human affairs.

During my early years, I played in the campus park directly adjacent to the building in which much of the Manhattan Project was taking place, although I only discovered that appalling coincidence many years later. At the age of eight (in 1940) I had a dream so horrifying I remember it vividly to this day. In a huge amphitheater, along with others, I was witness to an operation that took place, not in its pit, but high up along its periphery. A metal hatch slid open. A conveyor belt started delivering bodies of living children who were about to have their livers cut out. It appears that in the years 1939-1945, a number of psychiatric patients in Nazi Germany were having similar dreams (Charlotte Beradt. *The Third Reich of Dreams*).

My father seemed to find some kind of gratification in Hitler's program. At the age of ten, I remember telling him that a phenomenon like Hitler was possible only because of people like my father, and little old ladies who shared their genteel anti-Semitism over afternoon tea. What happened, I wonder, what happened to me in the intervening three years, so that at the age of 13, when the United States bombed Hiroshima and Nagasaki, somehow I was no longer present? In the context of such a devastating atrocity, had I managed to overlook the small matter of agency? For me, it was as if within those three years, history had taken on the kind of conveyor belt aspect of my own nightmare world. Events had somehow become mechanical, routine; as if there were no way of stopping—or opposing—a machine so given that questioning it was no longer relevant. The strange prescience of my childhood, in which I knew without knowing the unequivocal horrors of the crematoria, of experiments conducted on human flesh, gave way to a kind of numb acceptance, as if atrocity had somehow come to roost in my teenage room among the Teddy bears.

Somewhere in those three years, I must have lost the ability to feel; I must have succumbed to a deep ethical vacuum, a sloth beyond laziness, as if I had moved into a kind of dream world in which the most appalling juxtapositions seemed perfectly natural, and where there was utter suspension of disbelief. I suspect my own brutal awakening, sixty-five years later, might not have come about had I not accepted my dying neighbor's invitation to write what he called the Book of Life and Death.

There are times when one encounters a voice so deeply unsettling that it throws the assumptions of one's lifetime into question. Perhaps that is what people call a crisis of conscience, brought on, in my present case, by encountering John Gofman's curious little book, *An Irreverent Illustrated View of Nuclear Power.* His main point is that nuclear energy quite simply equals murder. From its first extraction from the earth in the form of uranium (and the radon gas the mining process releases), nuclear technology causes death, initially to the unfortunate miners who work in the uranium mines (for them, there may not be another choice) because work in the mines has a high probability of causing cancer or lung disease. On some level I knew that, certainly by the time in 1989, when with a friend, I traveled the Southwest. I remember a passing conversation in some deserted alley where outside her doorway a Pueblo woman was shaking out a rug. I must have asked her what work her husband did. He worked in the uranium mines. I asked her did he wear a respirator, use protective gear. She told me no. He wasn't given any safety equipment. I remember telling her she should make sure he had a respirator. So on some level I knew, but I failed to feel. I failed to question, I accepted uranium extraction as a given. Ultimately, I failed to make the quantum leap that as a subscriber, I used electricity that—in some part at least—*came from a process that accepted that people would have to die.* Besides the miners who extracted ore (and released radon into the atmosphere, consigning the planet to its ultimate death), there would be death attendant on the people milling the ore, refining and manufacturing fuel rods, nuclear plant workers consigned to working with highly lethal substances, ultimately affecting their health. I failed to understand that each year a nuclear plant operates, it essentially is bombing the earth by generating nuclear waste that can never be stored safely, that in one year, a 1000 megawatt plant generates waste the equivalent of 100 Hiroshima bombs, that in one year the 104 plants operating in continental U.S. are bombing the earth at least 10,400 times, and that the over 400 nuclear plants worldwide are bombing the earth 41,600 times, and that in the approximately

40 years after which their containment vessels become dangerously brittle, unable to withstand a serious challenge without failing (as we saw at Fukushima), the earth will have been bombed 1,640,000 times. That's over one and a half million nuclear bombs. And that, already in this half cycle of our nuclear existence, the energy half, it will have resulted in millions of deaths of cancer, and leukemias, not only of the working population that feeds this process, but also of the countless people who happen to live downwind where these plants routinely release plumes of radioactive steam, and where accidents ranging from relatively minor to major such as Three Mile Island, consistently occur.

And that is only half the Janus face of the nuclear cycle. The more obvious war-making cycle is the war-face, Gog to electricity's Magog.

Gofman's book was published at the height of the Cold War in 1979, while the Three Mile Island catastrophe was still ongoing. In those years, the public conversation centered around mutually assured destruction, the use of the Big Bombs, the ICBMs that could be fired from underground silos. It was taken as a given that if such weapons were made, they would eventually be used. In a way, it was like awaiting the Messiah. But what if the Messiah had already appeared in the smaller, under-the-radar guise of depleted uranium weapons? We will not have a nuclear war. *We are already having a nuclear war. Nuclear war is happening right now.* From the time the U.S. gave DU to Israel in 1982 for use in Lebanon, DU has contaminated—irreversibly and for all eternity—the soils of Kosovo, Iraq, (Gulf Wars I and II), Afghanistan, Pakistan, Somalia, Yemen, Libya, and wherever else the U.S. and its NATO surrogates have applied it. Why wait for the Big Bombs, when they can be dropped in the form of bite-sized pieces to even more devastating effect?

In my own gradual awakening, I have come at last to realize that for a great part of my life, I too have lived without a navel, without giving full weight to my own attachment to the earth, to the sanctity of all life on earth, and to the earth itself. It has been a ponderously slow awakening. Having gained some perspective at last on sixty-five years of my own deep moral abdication, I am left with a feeling of disgust at my failure to ask the questions that might have dispelled my own despicable vacancy an awful lot sooner. "There's no free lunch," writes

William Hendie: "if you want nuclear technology, some people have to die" (Gayle Green. *The Woman Who Knew Too Much*, 221).

THE TWO-HEADED MONSTER OF POISON FIRE

The nuclear weapons and nuclear power industries are two aspects of the same beast. Each exists in the presence of and as a result of the other. Every step of the nuclear chain contributes directly or through connecting steps to the virtually permanent contamination of our atmosphere, watersheds, soil and organic life.

Illustration by Mayumi Oda from *Safe Energy Handbook* by Jan Thomas, Claire Greensfelder and Wendy Oser (Berkeley: Plutonium Free Future, 1997)

99.
Where the 99% Comes In

Still in its infancy, the global Occupy Movement has emerged in response to a worldwide corporate agenda aided and abetted by governments riveted on keeping decision making an elitist, top-down affair. We may trivialize the Occupation by referring to them merely as encampments—although in this they keep fraternity with the Obamavilles of the have-nots, the Hoovervilles of our day—but much more essentially, by virtue of their daily "General Assemblies," they model themselves on the agora of antiquity, the public place of intersection where matters are discussed, and consensus is reached in a truly democratic way.

Today the Nashville agora, the Denver agora, and the Bangor agora are under police threat. Although the New York City's agora's generators have been confiscated by police, they have already been replaced by 14 power-generating bicycles. But so far this movement has responded to police crackdowns, notably in Oakland, with swelling numbers. Scott Olsen, the 24-year-old Marine with two Iraq tours of duty behind him, has lost (for the moment at least) his powers of speech, the result of a brain injury caused by our own domestic police violence. But for a movement which elevates free speech, his silence will amplify the movement's message: no more predation by corporations and their government front men, or the militarized goons who defend them.

In Tokyo, the Fukushima 100, a group of women from the exclusion zone, are sitting in at the Ministry of Economics, Trade and Industry, demanding a revision of Japanese energy policy. This is the time and place when the 40-year-old No Nukes movement and the OWS/99% movement must intersect. The nuclear industry, like Wall Street, conforms to the opaque, unaccountable corporate profile that has become the target of the 99%. Decisions were made to build and install plants by utility companies, in concert with corporations such as GE, Westinghouse and Babcock & Wilcox, with the skids being greased all along by an amply lobbied (bribed) government. At no time were people living within contamination radius of these plants seriously consulted, let alone warned, of any hazards involved. All risk-benefit decision-making that balanced health effects against economic and social benefit were "based on risk and benefit to society, i.e. government," and never factored in the cost to the individual or family that might have to absorb the pain of raising a handicapped child, or cancer and leukemia deaths, or, most grievous of all, loss of birthplace, home and countryside to nuclear contamination such as we saw in Chernobyl/Prypiat and are seeing now in the Fukushima exclusion zone (Rosalie Berrtell: *No Immediate Danger*, 54).

More than the ravages of disease, more than the weight of carrying offspring with physical or mental deficits, this loss of territory, with the pain and deracination that it involves, represents the ultimate genocide of a people. It is something I come close to understanding only in my nightmares. In some indigenous cultures, when an individual no longer speaks the language, he is thought to be dead. In others, when people lose their connection to the soil of their birthplace, they are said no longer to exist.

100.
"You Will Never Understand My Sadness…."

On the 230th day following the catastrophe, *The Japan Times* reports TEPCO has set up temporary offices and hired a staff of 1700 people to process scores of applications for compensation. Applicants are greeted by a letter from TEPCO's CEO apologizing for the inconvenience and anxiety caused by the accident. They are welcome to obtain compensation, but first they must fill out a 60-page application which may take up to 4 hours to complete—another inconvenience in a string of TEPCO-generated inconveniences such as admitting two months after the fact that three meltdowns had occurred, this by a CEO who eventually resigned after first trying to disappear.

Few applicants allow their feelings to break through their habitual reserve; occasionally a processing clerk will be privy to a confidence, but it is a confidence not meant to be shared. It will be the rare applicant who'll permit himself to murmur, "You will never understand my sadness." Rice farmer Naoto Matsumura is not one of them. Despite government orders to evacuate, he has refused to leave the exclusion zone. He lives in the abandoned village of

Tomioka, Fukushima Prefecture, amidst gardens grown tall with weeds, breathing the stench of dead cows left to rot in their stalls, and chicken coops alive with maggots. The only one left in a village that included 16,000 people, he tends the dogs and cats the villagers had to leave behind.

Although of samurai origins, and a former city dweller, Matsumura has become accustomed to the hardscrabble life of the countryside. "At first a few people stayed behind, some stayed with me in my house, but the last one left weeks ago. He asked me to take care of his cats." Without electricity and running water, Matsumura starts his generators up each night and draws his water from a local well. He survives mostly on canned foods, or fish he catches in a local, contaminated river. Once a month he runs his battered pickup past the police barricades to a distant town, to lay in supplies. The officers who are sent into Tomioka every day searching for looters and people who violate the exclusion order mostly ignore him. The official in charge of Tomioko's Living and Environment Division doesn't know he's there.

Matsumura admits he tried to leave once. He showed up at a relative's house, but she was so convinced that he was radioactive, she wouldn't let him in the door. He approached an evacuation center as a last resort, only to discover that it was full. The experience convinced him he must return.

"It's strange being alone at first," says Matsumura, but then he remembers the Japanese soldiers who refused to surrender until decades after the war ended. "I'm getting used to this life," he says:

> If I give up and leave, it's all over. It's my responsibility to stay, it's my right to be here. I've gone to Tokyo a couple of times to tell the politicians why I'm here. I tell them that it was an outrage how the cows were left to die, and how important it is for someone to tend to the family graves. They don't seem to hear me. We are already being forgotten. The rest of the country has moved on. They don't want to think about us.

101.
Russian Roulette

Day two hundred thirty-three following our nuclear catastrophe. Run by Entergy, in which the Clinton family holds a sizeable stake, Indian Point Energy Center, located at Buchanan, New York, on the Hudson River, 34 miles north of New York City, includes two working reactors, Units 2 and 3 (Unit 1 sits idle). Their licenses are due to expire in 2013 and 2015 respectively. As with Vermont Yankee, the GE Mark I BWR in Vernon, Vermont, another of its aging installations, Entergy is applying to extend the licenses of these reactors another 20 years.

In a 2008 article, based on evidence of 383 earthquakes in the New York region, seismologists at Columbia University's Lamont Doherty Earth Observatory show that an active seismic zone running from Stamford, Connecticut to Peekskill, New York passes less than a mile north of Indian Point. (A map of this fault line can be accessed at NRDC's report located at http://www.nrdc.org/nuclear/indianpoint/files/NRDC-1336_Indian_Point_FSr5.pdf. A target map centered at Buchanan, and various wind plume maps are also available at this location.)

In the paper cited above, the NRDC makes the case that Indian Point can be replaced at no immediate increase in consumer cost. In contrast to the disaster at Fukushima, an accident involving Indian Point's aging reactors would affect a densely populated urban area. Depending on the severity of such an accident, and on the direction of the wind, Entergy is playing nuclear roulette, minimally with anywhere from 2,800,000 to 9,900,000 people, not to mention some of the most-high priced real estate in the world.

Besides Manhattan, Long Island, Connecticut, and the densely populated cities of Newark and Paterson, depending on the severity of an accident and the wind direction, a wind-borne plume of radioactive debris could reach as far as Providence, Rhode Island, and Manchester, New Hampshire, above the Boston metropolitan area.

But Entergy's application process overlooks these risks—and overlooks its reactors' reference temperatures—that is, the temperatures at which aging reactors become so dangerously brittle they are subject to failure. Now that its licensing applications have come up for review,

Entergy has lost no time bringing yet another high-powered PR firm on board, the Breaux Lott Leadership Group, to euphemize nuclear issues. The two named partners are each former senators (Breaux and Lott). Their sons are junior partners, guaranteeing the Entergy account stays in the family.

Job seekers eager to enter the PR market might benefit from studying some of the tricks of the PR trade quoted below:

> Indian Point Energy Center: Safe. Secure. Vital. N.Y.C. report finds cost and pollution will rise if Indian Point is closed.
>
> Entergy and Indian Point Energy Centers have a tradition of deep and lasting corporate contributions...generously contributing to non-profit organizations, schools and Universities in New York. Specific initiatives include the food pantry at Zion Episcopal Church...
>
> Carbon Disclosure Project Salutes Entergy's Approach to Climate Change.

NRDC's report includes maps of wind "rose" and wind "petal" simulations. The tide of one of these "petals" washes over Fire Island, the 31-mile-long sand bar lying out to sea, protecting the Great South Bay and the Long Island shore.

102.

Day 276. NRC Announces It Has Developed Slow-Motion Technology

Luntz Global™ is a powerhouse in the profession of message creation and image management.

We have counseled Presidents and Prime Ministers, Fortune 100 CEOs and Hollywood creative teams in harnessing the power of language and visuals to change hearts, change minds and change behaviors. We have become a hyper-

attentive nation that is quick to judge. The words and visuals you use are more important than ever in determining whether you win or lose at the ballot box, the checkout line, and the court of public opinion. We know the words that work. Do you?

Our confidence comes from decades of research, polling, and consulting to the opinion elite worldwide, with proven results that withstand the test of time.

Remember: "It's not what you say. It's what people want to hear."

Frank Luntz is the principal of Luntz Global, a man with a wide reach, and author of the bestselling *What Americans Want—Really*. He has put words in the mouth of every two-bit snake oil salesman who's ever run for office. But on the subject of nuclear energy, he waxes particularly eloquent. "Americans aren't concerned about nuclear energy since Fukushima," he counsels members of the industry, "they are [merely] anxious." And here's the good news from Frank: "their anxiety can be managed, controlled and addressed," (but presumably not in that order). All you have to remember when talking about the absolute beauty of nuclear power is that although uncertainty has grown since Fukushima, "the light (for the industry) is yellow, not red." They need to hear you're "learning lessons from Japan," in every way, and every day. And "going forward" (a red-flag clue a bromide is about to be delivered) there's going to be "even more accountability and even better safety." All around. Your responsibility is to educate them (if that is indeed possible), by reminding them that you are "committed to the relentless pursuit of safer nuclear energy." Let them know they "have a right to know the FACTS about nuclear energy…and that [you] have the responsibility to tell [them] openly and honestly."

But "talk about radiation as little as possible." Or if you must talk about it, say "we are committed to safely containing radiation in every nuclear facility. We don't take chances. We use layer upon layer of redundant protection." And above all, "you need to invest time and language in building up the stringent oversight of the NRC." Frank follows up this generous advice with a list of "Do not says" and "Do says." Example: "never refer to the NRC as a government agency." Say, "the NRC, the independent watchdog regulator."

Gregory Jaczko, the U.S. top NRC regulator, seemed to be borrowing from Frank Luntz's down/play book on Day 276 when, in an interview at Bloomberg's headquarters in New York, he opined that the New York City area could be safely evacuated in the event of a Fukushima-like disaster at Indian Point nuclear plant located 34 miles to the north, because a crisis would unfold slowly (Brian Wingfield and Julie Johnsson, "New York Nuclear-Accident Evacuation Would Work, Jaczko Says," *Bloomberg Businessweek*, January 19, 2012).

"Nuclear accidents do develop slowly, they do develop over time, and we saw [sic] that at Fukushima," Jaczko said. "It's unlikely a nuclear accident would require prompt action beyond more than a few miles, where the highest radiation levels would be."

It might be safer to say that what developed slowly at Fukushima was any effort on TEPCO's part to disclose what was actually happening there until long after the fact, and on the part of the Japanese government, which kept as tight-lipped as possible as long as possible, and continues to this day delaying the evacuation of women and children who are already showing signs of radiation poisoning. Jaczko seems to be unaware that exactly at 3:29 p.m., *minutes* after the earthquake struck that destroyed the GE Mark I BWRs at Fukushima—and long before the first tsunami wave rolled in—the first radiation alarm went off.

But perhaps, unbeknownst to the rest of us, the NRC has developed slow motion technology such that it can decelerate catastrophes by a factor of thousands with special application to Indian Point.

On December 3, TEPCO issued the summary of an internal investigation, led by executives and reviewed by a panel of external experts picked by the company, and based its findings on plant data and hearings conducted over the past six months.

- It hasn't found the source of irradiated-water leaks.
- It doesn't know why radiation releases spiked on March 15.
- It says the tsunami, not the quake, caused plant damage.
- It finds no evidence of big errors in its response.

(Mitsuru Obe. "No Error in Nuclear Crisis," *The New York Times,* December 3, 2011).

103.
Through the Links

Fire Island on the Atlantic seaboard piqued my imagination from the time my mother told me how Grandma Bazinet, whom we used to see in the notions shop where we went to buy thread, was swept out to sea in the hurricane of 1938 inside her Fire Island house. For someone like me, raised in the steep canyons of New York City, used to the dismal asphalt of its streets, where the sidewalk trees—if there were any trees at all—were caged like zoo animals inside metal grilles, Fire Island was where I first learned there are places in the world where you can see forever.

I remember the house where I summered, its magical roller piano, its screened-in upstairs porch. Many years later that house had become hidden behind a dense row of hedges. I tried in vain to find it—until the particular smell of pine resin led me to the place. I remember the first time I saw fireflies at night, the grit of sand on the raised wooden walkways, and the intoxicating scent of pine resin in the noontime heat of summer, the sound of cicadas drumming in the fall. I remember walking the beach for miles, the feel of hardened sand underfoot, singing at top voice with no one to hear but the wind and the waves, the swell of the dune line, sutured now and again with rolled slat fencing, the dune grass riffled by strong winds. The sighing and bubbling of the tide. My first sexual awakening. Playing tag, dancing with the waves. The smell of the sea and sand. The tiny sand crabs digging frenziedly as the retreating arcs of water exposed their breathing holes. Walking alone at sunset, reaching up to grab onto a buoy rope and being swept out to sea, loosening my grip to fall—15 feet—just short of the incoming waves.

And many years later, returning before dawn to watch the sun rise over the Atlantic, and discovering the golden sands studded with shining white, sea-smoothed stones of alabaster—thousands of them. But beyond all remembered treasure, it's the sweep of the shoreline in the lost distance that still makes me want to celebrate the gathering of the waters as pulse of wave breaks once again, over and over, to the limit of the eyes, in this place where the earth first revealed to me the grand sweep of her curvature.

But what if....What if Indian Point blew? And what if a radioactive plume from Indian Point swept over Fire Island? What if the wind petal engulfed this place, a place like no other? What then? Could I picture it? Police cordons, roadblocks, spike strips? Can I hear the sirens? Can I see the guard posts, the military police, the radiation warnings, the chain link fencing—miles and miles of it—and no way in the space between the links—ever—to see my world made whole again.

Which place in the world is most marvelous to you? If, say, it were sacrificed in the interests of the nuclear industry for 4.5 billion years or all eternity—whichever comes first—how would you feel?

104.
A Little Bit Goes a Long Way...

Television sets, refrigerators, milk cans, recliners, tables, flatware, china, étageres, bathroom fixtures, kitchenware, hi-fis, CD players, VCRs, dressers, nightstands, tables, drawing boards, books, fish tanks, records, CDs, samovars: pass them by. If you go shopping in the flea markets anywhere in the million square miles of Belarus or the Ukraine, pass them by, no matter how hot the bargain.

If you listen to Sergei Vasilyevich Sobolev, Deputy Head of the Executive Committee of the Shield of Chernobyl Association and curator of the Chernobyl Museum, you'll discover where they came from. Hoping to escort an English photographer through the exclusion zone, he was getting passed from one apparatchik to the next, all of them making creative excuses: no maps exist; there's no way to get permission, etc. etc. The photographer was looking to take pictures of all the burial trenches he'd examined from the air: places where the red forest is buried (from when the trees turned red); places where he had seen "thousands of pieces of automotive and aviation machinery, fire trucks, ambulances—the biggest graveyards right next to the reactor." One day it dawned on Sobolev: of course! the graveyard doesn't exist. All the stuff's been stolen, "carried to market, for spare parts, for collective farms, and people's homes." It's on the account books, all right, but it's a graveyard that's risen from the dead.

Little bits of radiation go a long way. Immediately following the explosion and fire, three hundred forty thousand young Russian soldiers were called up to clean Chernobyl of graphite and radioactive waste. All of them died of radioactive poisoning. No record remains of their names, who they were, where they were born or died, or of their cause of death. Of these, the three thousand six hundred who worked on the reactor roof with nothing to protect their legs and feet but imitation leather boots died first. They were called "green robots" because they replaced all the robotic clearing machinery that failed in conditions of extreme radiation.

Four hundred miners came from Moscow and Dnepropetrovsk tunneling beneath the reactor to freeze the earth with liquid nitrogen so concrete could be poured for the sarcophagus. They worked on their knees, half-naked in extreme temperatures, pushing mining carts before them. None survived.

> There was a moment when there existed the danger of a nuclear explosion, and they had to get the water out from under the reactor so that a mixture of uranium and graphite wouldn't get into it—with the water they would have formed a critical mass. The explosion would have been between 3 and 5 megatons. This would have meant that not only Kiev and Minsk, but a large part of Europe would have become uninhabitable....So there was this task: who would dive in there and open the bolt on the safety valve...? The boys dove, many times, and they opened that bolt.... These people don't exist anymore, just the documents in our museum with [their] names (Sergei Vasilyevich Sobolev, quoted in *Voices from Chernobyl,* 133).

The slight confusion clouding El Presidente's reassuring 2009 Prague statement, "To put an end to Cold War thinking, we will reduce the role of nuclear weapons in our national security strategy....

[but] make no mistake: the United States will maintain a safe, secure, and effective arsenal to deter any adversary" has been clarified by a FY2012 defense budget. Totaling some 213 billion, it includes, among other items, funding for:

> three new nuclear weapons productions plants; a new plutonium 'pit' manufacturing facility at the Los Alamos Lab in New Mexico; a production facility for uranium components under construction at the Oak Ridge National Laboratory in Tennessee; and a replacement for the Kansas City plants in Missouri, which manufactures most non-nuclear nuclear weapon components.... Of that, 3.7 billion is earmarked to be spent developing a new Air Force long-range bomber with nuclear capabilities (Jackie Cabasso, address delivered on Hiroshima/Nagasaki Day, 2011, at the gate of the Lawrence Livermore National Laboratory, California).

There may be some question as to why bombers and weapons are still needed when nuclear plant explosions and climate collapse are doing quite an admirable job of decommissioning the planet and causing untold human suffering, but then, when it comes to military spending, parsimony has never been Washington's strong suit.

"When I was a child, I used to listen to stories about a man who went over Niagara Falls in a barrel. I thought what a fool! Only a fool would choose to go over Niagara Falls in a barrel. The way I see it," my hiking partner says, "we're in some kind of boat. But we've lost the oars and the boat is caught in the eddies, slowly spinning round and round. The current is taking us nearer and nearer to the edge of the falls."

105.

Losing Ground...

Contrary to TEPCOs sunny reassurances, things at Fukushima are not stabilizing quite yet. On the 236th day following the first explosions at Fukushima, xenon-133 and 135 have been detected in a filter at Fukushima GE Mark I Unit 2, an indication, because of xenon's half-life, that fission is still occurring. An independent study has effectively demolished TEPCO's and the Japanese government's carefully managed minimalist scenario. France's l'Institut de Radioprotection et de Sureté Nucleaire (or IRSN) has issued a recent report stating that the amount of radioactive cesium-137 that entered the Pacific after March 11th was probably nearly 30 times the amount stated by Tokyo Electric Power Co. in May. According to IRSN, the amount of the radioactive isotope cesium-137 that flowed into the ocean from the Fukushima Daiichi nuclear plant between March 21 and mid-July reached an estimated 27.1 quadrillion becquerels. Radioactive cesium-137 has a half-life of roughly 30 years, so if the IRSN estimates are accurate, then by 2041 the Pacific's aquatic life will only be subjected to 13.55 quadrillion becquerels of radiation.

Likewise, at Chernobyl, in 1986, the Soviet government resorted to every possible delay to avoid creating panic; because of stalling by party bosses and government officials, thousands of people were needlessly exposed, not only those who were called up as liquidators and given no information or appropriate protection, but people who made their lives in surrounding villages, who sunbathed on the grass, and who swam in the rivers, unaware that their entire world had become an uninhabitable zone.

> For…Belarus…during World War II, the Nazis destroyed 619…villages along with their inhabitants. As a result of Chernobyl, the country lost 485 villages and settlements. Of these, 70 have been forever buried underground. During the war, one out of every four Belarussians was killed: today one of every five Belarussians lives on contaminated land.., 2.1 million people, of whom 700,000 are children" (from "Chernobyl" *Belaruskaya entsiklopedia* quoted in *Voices from Chenobyl,* 2).

106.

Recording the Future…

In *Voices from Chernobyl,* Svetlana Alexievich reveals life in the zone following the accident. She lays bare the dimension of human suffering experienced by the thousands of people she spoke with over the three-year period during which she conducted interviews. Alexievich is a professional journalist by training, but here she draws attention to her problem with fact-telling:

> I often thought that the simple fact, the mechanical fact, is no closer to the truth than a vague feeling, rumor, vision. Why repeat the facts—they cover up our feelings. The development of these feelings, the spilling of these feelings past the facts, is what fascinated me.… These people had already seen what for everyone else is still unknown. I felt like I was recording the future.

Children's Chorus

> The sparrows disappeared from our town in the first year after the accident. They were lying around everywhere—in the yards, on the asphalt. They'd be raked up and taken away in the containers with the leaves. They didn't let people burn the leaves that year because they were radioactive, so they buried the leaves.
>
> The sparrows came back two years later. We were so happy, we were calling to each other: "I saw a sparrow yesterday! They're back!"
>
> The May bugs also disappeared, and they haven't come back. Maybe they'll come back in a hundred years or a thousand. That's what our teacher says. I won't see them.
>
> September first, the first day of school, and there wasn't a single flower. The flowers were radioactive. Before the beginning of the year, the people working weren't masons, like before, but soldiers. They mowed the flowers, took off

the earth and took it away somewhere in cars with trailers.

In a year they evacuated all of us and buried the village.... First they tear a big pit in the ground, five meters deep. Then the firemen would come up and use their hoses to wash the house from its roof to its foundation, so that no radioactive dust gets kicked up. They wash the windows, the roof, the door, all of it. Then a crane drags the house from its spot and puts it down into the pit. There's dolls and books and cans all scattered around. The excavator picks them up. Then it covers everything with sand and clay, leveling it. And then, instead of a village, you have an empty field" (*Voices from Chernobyl,* 218–219).

107.
In America There is No Death...Only *The New Yorker*

On the 220th day following the Fukushima Daiichi planetary disaster, *The New Yorker* magazine publishes a "Letter from Fukushima." The writer, Evan Osnos, now a *New Yorker* staff reporter, vaunts a resume that includes Harvard, and *The Chicago Tribune,* and the Asia Society Elliott Osborne Journalism Prize. Our golden boy has written a lovely treatment of a disaster, aestheticized to harmonize with the New Yorker's Tiffany ads, replete with tasteful photographs and reassuring statements: "Opponents [of nuclear power] cannot easily suggest that a meltdown will produce the huge number of immediate casualties that the public imagines." He quotes Haruki Mirakami, perhaps Japan's most touted writer: "'This [Fukushima] is a historic experience for us Japanese: our second massive nuclear disaster. But this time no one dropped a bomb on us. We set the stage, we committed the crime with our own hands,'" in a nice example of self-flagellation and blaming the victim. He rivals Yukio Edano for airbrushing with his assertion that the Fukushima meltdowns "scattered nuclear fallout over an area the size of Chicago." All of which seems to suggest that journalism prizes are won by people who don't ask the uncomfortable questions and who promise to uphold the comforting narratives. On the 240th day of our planetary disaster, I respond with a letter:

Editor
The Mail
New Yorker Magazine
themail@newyorker.com

Evan Osnos omits some very essential facts. By foregrounding the tsunami damage to Units 1, 2, and 3, he avoids mentioning containment failure. Long before the tsunami hit, radiation warning sirens went off. As early as late March, in a report called "Reactor Core Status of Fukushima Daiichi Nuclear Power Station Unit One," TEPCO admitted there was pre-tsunami damage to key facilities, including pipes.

As far as I know, whatever his talents, Murakami is not a nuclear authority. The question to ask is: who designed these reactors? Who installed them? All the Fukushima Daiichi rectors are GE Mark I boiling water reactors. They were sold to Japan knowing they were defective. Three of GE's design engineers quit rather than collaborate on a design they believed would fail. One of them, Dale Bridenbaugh, called the Mark I BWR ten pounds of energy in a five-pound sack. The Fukushima 20-kilometer exclusion zone affects a rural, sparsely populated area. Two hundred thousand people lost their land and their way of life. But there are 23 GE Mark I BWRs still operating in the United States. If there were a similar accident at one of them, Vermont Yankee, operated by Entergy, millions of U.S. citizens would be irradiated, and would have to lose their homes, and their land would become an uninhabitable zone for millions of years.

> The Fukushima fallout is contaminating an area larger than the size of Chicago. According to the Centers for Disease Control, in 8 northwestern cities in the United States, in the first three months following the events of March 11, infant mortality rose 35%.

The odds are long they'll print it, and even if they do, how many of its more than one million readers will notice a small letter on page 7 before turning to more responsible media reportage?

108.
Visit to the Consulate...

The morning dawns bright with autumn sunlight on the 241st day of our ongoing planetary catastrophe. A set of coincidences finds me pacing the pavement alongside 50 Front Street in San Francisco, where the Japanese Consulate keeps its offices. I display a home-made sign:

C.D.C.:
35% SPIKE
U.S. INFANT MORTALITY
AFTER
FUKU-FALLOUT
www.stuartsmith.com

Most passersby are on their way to work. They cut their eyes sideways, or avoid looking altogether. Few people manage to really read the message.

The call is for 9 am, and although the others begin to show up around 9:20, all take the events at Fukushima very seriously. Many of the women have children. One of them has demonstrated for a number of years at the Nevada test site; another with her documentarian husband is the founder of The Ecological Options Network (www.

eon3.net). EON has banded together with Beyond Nuclear, Nuclear Information and Resource Service, and Citizens for Health to form FFAN, the Fukushima Fallout Awareness Network. We are joined by skateboarding Kafyr, who's on the staff at Greenpeace. We're here to offer the Japanese consul a petition signed by over 6,000 people in support of the mothers of Japan. It calls for immediate halt to the incineration and shipping all around Japan of millions of tons of radioactive rubble and debris from the Fukushima Daiichi disaster. The full text of the petition reads:

> We are writing this letter in support of...thousands of mothers across Japan....We believe the government's negligence will have more adverse consequences than the already catastrophic impact of the tsunami and resulting radiation exposure. An almost certain rise in cancer rates for millions of people is the best case scenario from the continued leakage from Fukushima Daiichi. It is our intention to limit the exposure of human beings to this risk to the greatest extent possible.
>
> It is the belief of the undersigned that the dangerous radioactive rubble at Fukushima Power Plant and the other areas around must be left at the site of the disaster. Efforts must be focused on ending the ongoing fires at the plant, and people should be evacuated from the immediate area in accordance with radiation levels set before March 11th. All recent Japanese Government policy changes to increase allowable radiation levels must be overturned to pre-disaster levels.
>
> Today Japanese government is systematically spreading radioactive material and publicly hosting events to eat food from Fukushima as a patriotic act, raising radiation safe standard[s] for food and rubble alike. For example in Japan today food reading 499 bq/kg can be legally distributed in the market without any label for consumers. [It] has twice raised allowable levels of radiation for rubble which they will now ship across the country to be burned and dumped into the ocean at [various] locations including Tokyo Bay.

This negligent behavior must be stopped or an already devastating event will turn into an historic environmental disaster [of] international reach.

If the rubble piled up everywhere were not a big enough problem for the government, there is the added fact that much of this rubble contains radioactive material from the nuclear spill.

Tokyo's local government officially accepted 1,000 tons of rubble from Iwate. Starting at the end of October, 2011, they will transport the debris on trains and burn it and use the ashes as landfill in Tokyo Bay. Iwate Prefectural government estimates indicate that the rubble contains 133 bq/kg of radioactive material. This would have been illegal before March but the Japanese Government changed the safety level for rubble from 100 bq/kg to 8000 bq/kg in July, 2011, then again to 10,000 bq/kg in October. Tokyo officials announced that they will accept 500,000 tons of rubble in total.

The problem is not restricted to the Tokyo area, which is geographically near the impacted areas. The governor of Tokyo stated that he hopes this would encourage other local authorities to accept rubble. It is difficult to accurately [gauge] the consequences of these government actions, but no one can [deny] that a huge environmental gamble is being waged. If many other local governments in Japan decide to follow Tokyo's lead it will cause areas [which] are not yet directly impacted by the radioactive spill to contaminate their local soil and water.

The Minister of the Environment, Mr. Hosono, said in a September 4, 2011 press conference that "it is the consideration of the national government...to share the pain of Fukushima with everyone [or everywhere] in Japan," reiterating his intention to create a final processing facility outside Fukushima Prefecture for debris and dirt from near the nuclear accident to be burned.

We are asking you to please discourage the Japanese government from spreading, burning and dumping rubble

> from contaminated areas. It should be left on site and people should be evacuated from those areas according to the standards in place before March 11th. It is the opinion of the undersigned that, if allowed to proceed, we will witness an historic error conducted by the Japanese government that will negatively impact human lives for hundreds of years. The alternative is that we act immediately to prevent this unnecessary outcome, and history will remember this only as the time that Fukushima Daiichi region was rendered uninhabitable rather than a worse, if uncertain, alternative.

Before our meeting with the consul can take place, there is a minuet of formalities, formalities initiated—very courteously—by the Japanese staff people; formalities initiated by the building receptionist who demands to duplicate our driver licenses; formalities on the 23rd floor, where one by one, we must pass through the metal detectors.

We file in to the smallest conference room possible, where the consul greets us, accompanied by a stony-faced assistant whose expression never changes throughout the meeting. We present the consul with a bouquet of yellow sunflowers and roses, and our petition in English and Japanese wrapped in yellow tissue paper to which a rose has been attached. One by one, we state our case. We speak of the children of Japan, some of whom are being challenged to drink contaminated milk to show their patriotic loyalty, many of whom are being needlessly exposed to radiation. We speak of our desire to protect our own children from the steady stream of Fukushima fallout blanketing the United States. We implore him to convey to his government our concern that nuclear waste not be scattered over Japan and disposed of by burning. And we apologize. We apologize for Hiroshima. We apologize for Nagasaki. We apologize for all the harm our country did to the people of Japan. And we accuse General Electric, which sold the Mark I boiling water reactor to Japan—and to utilities all over the United States—knowing it was defective.

It is the consul's turn to respond. We wait expectantly. We watch a man, an official, break down in tears. He wants to speak, but he cannot. We watch as he removes his glasses and unfolds his carefully folded handkerchief to wipe his eyes.

109.

A 30-Foot Hairline Crack and Tiny Little Offspring Cracks…

There are many ways to tell this story. Let's start with the scientists. On the 242nd day following our planetary catastrophe, on November 8, the Union of Concerned Scientists sends a letter to the NRC drawing attention to a 30-foot (that's three stories long) hairline crack in the containment vessel of the Davis-Besse nuclear plant outside Toledo, Ohio (population 287,000). In 2002, Davis-Besse nearly melted down, causing a long shut-down. It was not David-Besse's first close call or protracted shut-down.

First discovered by contractors from the Akron Company hired by FirstEnergy to cut through the 2.5-foot-thick wall of the barrier building that shields the reactor, and through the containment vessel in order to replace the rusted reactor head, engineers found the 30-foot hairline crack. Further examination revealed numerous other tiny cracks on the façade of the containment building, which in turn houses the steel vessel containing the reactor. To rule out a wider problem, Akron Company hired structural and concrete engineers to determine if there is internal cracking. Sonogram tests found numerous tiny sub-surface cracks in the building's external façade. Engineers took core samples from the external façades and the walls themselves, a total of 50 core bores up to 24 inches deep, and 11 samples of the building's walls. Inspectors found cracking in the main walls of the building in two areas where steam pipes pass out of the barrier building and into the turbine and generator building. Their analysis showed that the cracks were related to the nearby façade.

When aging reactors receive enough radiation bombardment over the forty years or so they were built to operate, their concrete walls become brittle and subject to failure.

Imagine now that Davis-Besse had not needed to have its reactor head replaced. Assume that no hole had needed to be cut. Assume that no workers had erected the huge scaffolding necessary to reach the top of the reactor. Would anyone have noticed any hairline crack, let alone one 30 feet long? Imagine that a routine inspection of the plant had been conducted by the

Nuclear Regulatory Commission, in whose interest it is to continue operating such a reactor without drawing attention to any recently appearing flaws. What is the likelihood that such a defect would have been discovered in time to prevent catastrophic failure in the event of an accident?

Why in this story do I find this contrast of social class, occupation, vested interest, and hierarchy so telling? Who is more inclined to take local interests to heart? And why? And who will determine when the 103 other aging reactors scattered throughout the United States have begun to show signs of reaching their brittle stage in time to prevent what could become another Fukushima?

Most of the 104 reactors located throughout the United States are reaching the end of their useful lives. A determination is still to be made: will Davis Besse ever operate again? Tune in next week.

110.
Breaking Down, Linking Up

On this, the two hundred forty-ninth day following our planetary catastrophe, agents of the state have emptied Zuccotti Park, they have emptied Oscar Grant Plaza, they have emptied Sproul Plaza at the University of California. This is a concerted effort, evidenced by an admission by Jean Quan, agrammatical Mayor of Oakland, that now cities as well as corporations are human beings: "I was recently on a conference call with 18 cities across the country who had the same situation. . . ." She was speaking in an interview with the BBC before a wave of raids broke up Occupy Wall Street encampments across the United States. Days later, in response to a Freedom of Information Request, the FBI, the CIA, and Homeland Security came up empty handed, unable to find any documentation of a concerted national effort to clear the 99% campgrounds! No matter. Doggedly people will return, will occupy again, have returned, have occupied again. No place on earth can be de-peopled very long (not with 7 billion and procreating). Even Japan, where today, it has been determined that the entire prefecture of Fukushima has become too hot for human habitation.

Speaking of the Occupy Wall Street movement, Harvey Wasserman writes in *Common Dreams* of October 24:

> Through the astonishing power of creative non-violence, [the Occupy Movement] has the magic and moxie to defeat the failing forces of corporate greed. Such a moment must come now in the nick of time, when the corporate ways of greed and violence pitch us to the precipice of self-extinction..... Atomic energy has failed on all fronts: Once sold as too cheap to meter, it's now the world's most expensive generator. Embraced as a corporate bonanza, it can't obtain private liability insurance [or] private investment. Worshipped as a technology of genius, it cannot clean up its own radioactive messes. Described as the 'magic bullet' that could power the earth, it's now the lethal technology threatening to destroy it....The key is to deny the [U.S.] nuclear industry the federal funding without which it can't build [new] reactors. And here's where the Occupy and No Nukes movements intersect. If we can hold off loan guarantees for another year or two, and shut [down] some older reactors like Vermont Yankee and Indian Point, the dam will break, and the corporate impetus to build new rectors may finally [end.] (Harvey Wasserman. "Where Occupy and No Nukes Merge and Win").

Fifty-three days after Harvey Wasserman publishes his prescient words, the NRC holds a public hearing in Michigan where it presents a 1500-page EIS statement that disregards the 16% reduction in Michigan's use of electricity over the past ten years and claims that a new Fermi 3 nuclear reactor proposed by Detroit Edison would have no adverse environmental impact.

Occupy Toledo crosses the border into Monroe, Michigan to confront the NRC at its own hearing. Amplified by Sierrans and other environmentalists, they mic-check the panel: "We denounce this ridiculous public meeting as nothing but a dog and pony show...and hold you criminally responsible for prioritizing the profits of Detroit Edison over the health and safety of the citizens and the natural environment of this region. We will do everything in our power to stop this plant from being built. We are the 99%."

The commissioners look on in stony silence as the facilitator tries to shout the protestors down, until he is forced at last to suspend the meeting.

Occupy Toledo sees this as an exhilarating and empowering start with the potential to resonate across the country.

111.
Four "No Problems"

Investigative reporter Greg Palast's latest book, was overdue. How could he keep to his publisher's deadline and still report on Fukushima, the greatest nuclear accident the world has ever seen? Palast works fast, aided by timely little packages that keep dropping through his transom just when he needs them most. Behind his carefully cultivated persona (battered fedora, battered gonads) lies a reporter with the savvy of a snake. But for the grace of god and a Jewish father (who liked to say: "you have to have a lion's heart and a steeled soul to face a world too terrible for words"), Palast himself would be feasting at the vulture's picnic, the title of his just published book. There's grief in it for every taste, but I zero in on what he has to say about the nuclear industry, the daisy chain of utility CEOs, the NRC, and the sorry politicians caught with their pants down, all playing nuclear roulette with our lives, our landscape, and entire territories the planet over in a devil's tango of greed unmatched in human history, unmatched simply because, until Geo. Herbert Walker Bush trumpeted the New World Order, nothing on this globalized economic scale had ever been imagined.

Palast, who has advised regulators in 26 states and 12 nations on the regulation of the utility industry, plays his hunches. He shares the subterranean mentality of his quarry: no bottom feeding is too slimy for his imagination. But the lizards of the nuclear industry are particularly impervious to scrutiny. Get the drop on NINA (Nuclear Innovation North America), and under the bra, you get NRG (like in En-er-gy); drop the pasties off NRG, you get Reliant (another of Obama's favorite girlies), under Reliant's g-string you get Brown & Root—Halliburton's g-spot. But you're still coming up short because a company currently going by the name of Shaw (née Stone & Webster) will build all the new U.S. reactors, and new ones in the UK as well. And Shaw is right there, part of NINA's toxic consortium, along with NRG, Reliant, and Houston, and keeping company with a Japanese hottie more commonly named Tokyo Electric Power Co., that will get 20% of the pie and a 12% ownership slice. A girlie show is just the ticket when you're trying to hide fraud.

In 1989 Palast was gumshoeing on a case brought against the Shoreham nuclear plant located in Suffolk County, New York. His files in connection with that case allow him to navigate the complexities of the nuclear industry today, with particular attention to Fukushima Daiichi.

Palast's files yield a facsimile copy of Gordon Dick's field engineer's log book, which refers to **Seismic Qualification** tests conducted at Shoreham nuclear plant. No reactor anywhere can run without a certified SQ. Dick ordered the man who ran the SQ test to change his report to the Nuclear Regulatory Commission from "failed" to "passed." But his subordinate, a man appropriately named Wiesel, felt conflicted. To file an accurate report with the NRC would mean he'd lose his job; to falsify his report was (and is) a federal crime. He filed a fake report. It took one of Palast's assistants four months to comb through millions of pages of documentation—most of it handwritten—in the Shoreham nuclear plant's file room at the NRC to find it. The name of the construction company: Stone & Webster Engineering, which built one third of the reactors in the United States, and now masquerades under its new name, Shaw Construction.

Which leads Palast to question whether TEPCO may also have fudged Fukushima Daiichi's seismic qualifications. TEPCO itself claimed that Reactor 2 was hit by a seismic force of 550 gals (as in galileos, a unit of ground movement). But Unit 2 was designed to cope with only 436 gals. Palast claims that documents show that TEPCO fooled Japanese government regulators by never delivering on its 2006 promise to raise seismic hardening to 600 gals, proof that TEPCO was more worried about its bottom line than bothering with seismic safety.

Why didn't the emergency diesel backup keep Fukushima's reactor cores from melting? Palast suspects that Fukushima's diesels—like most other diesel backups worldwide—were never expected to work, that "they can't work. Not anywhere: not in Japan, not in the Untied States, not in Russia, nowhere." In an emergency, **diesel backup generators** are supposed to crash-start, revving from zero up to 4,000 rpm in ten seconds or less. Palast's 1986 files yield a page from Emergency Diesel expert R.D. Jacobs' notebook. Jacobs was hired to test Shoreham's backup cooling system. The record states that on his last visit he pressed a company executive to open the motors and have them inspected because he couldn't tell what the axial vibration of the crankshaft was doing to the diesel units. Tested for a few minutes at low power, they were fine, but to "crash start" them, Jacobs had his doubts. The power company management told him to get lost. Later, testifying in court against the Shoreham plant, Palast writes: "When we test[ed] the three Emergency Diesel generators in emergency conditions, one failed almost immediately (the crankshaft snapped), then the second, then the third. We named the three diesels 'Snap. Crackle, and Pop'" (Palast, *Vultures' Picnic*, 294). "Every nuclear plant in operation today, and the ones they want to build, all depend on this same emergency diesel engine set-up to save your behind from a nuclear meltdown. Good luck" (*Vultures' Picnic,* 287).

Once the Shoreham diesels were shown to fail, the Nuclear Regulatory Commission found several other plants with inadequate diesel backup, especially diesels made by Transamerica Delaval Inc., but with a little Reagan-era manipulation, it turned out that fixing the regulators was cheaper than fixing the problem. After all, to replace them with appropriate equipment would cost a billion dollars per unit. It's all about those rems for rads.

Palast cites recent evidence from an engineer who worked for General Electric, the same GE that built the Mark I BWRs at Fukushima (and 23 still operating reactors scattered all over the United States), who claims "the one very large vulnerability that the system had was flood." His original warning came in 1985, but to have **water-sealed** the already existing diesel backup buildings at Fukushima would have cost TEPCO billions. It didn't help that in preparing the ground for the plant's foundations, 30 meters (nearly 100 feet) were shaved off the hill before the reactors were installed. Diesels vulnerable to water remain standard industry practice worldwide (except in Germany). Palast suspects that at least two of the Fukushima diesels most likely failed even before the tsunami hit.

One of Palast's sources, pseudonymously identified as Feuerman (he prefers to keep his job), is a nuclear fire suppression specialist, who "directed tests for **fire safety standards** used at every reactor worldwide." But the standards provided by the International Atomic Energy Agency switched false data in place of Feuerman's so that reactor fires would appear less of a threat, and with these nice faked standards, IAEA got the industry off the hook. "This is the nuclear industry's Safety Showtime. The failure is in the design, the design of the political system, the corporate system.... Fraud is as much a part of the structure of a nuclear plant as the cement and steel" (*Vultures' Picnic*, 305).

112.
NRG and Southern: Evil Twins

Prior to awarding its new $56 billion nuclear plant contracts, the Obama administration may have preferred not to conduct appropriate background checks on Southern and on NINA (formerly NRG), but in the absence of such checks, I thought I'd try to simplify NINA's tangled organizational charts to make it easier for the GAO.

NRG low-bid Obama's new nukes contract at 5.709 billion—except that NRG's internal documents show various other figures, among them a cost of 14.272 billion; that's 7.1 billion per reactor, or a 23% cost overrun—and the project hasn't even started! Creative accounting yes, and in normal circumstances it would qualify as fraud but not when there's corporate/government collusion. Not happy with dreaming up the 23% hike, the cost was upped another 5 billion to 12 billion by Toshiba (which bought out Westinghouse). It just turns out that Toshiba is really Shaw Construction—one of Toshiba's major stakeholders—and the same company—under its former name of Stone & Webster Engineering—made famous for its diesel popping, seismic faking safety concerns and for conviction, along with Long Island Lighting Company for conspiracy under the RICO (racketeering) Act (*Vultures' Picnic,* 302–304). No doubt under their new NINA avatar, Toshiba and NRG will both benefit from Shaw Construction's creative billing systems.

> *Shaw has been a pioneer in nuclear plant design and construction for more than 60 years. Backed by experience and poised to meet the growing demands of the future, we continue that legacy today. From our ongoing work in China to our contracts in the U.S., Shaw is leading the industry's global nuclear resurgence.*

(How much did they pay Frank Luntz, I wonder, to generate that pabulum?) Shaw's PR doesn't let on that Shaw holds a major stake in Toshiba. It also doesn't happen to mention that, in its original avatar of Stone & Webster Engineering, it filed fraudulent seismic qualification and backup systems reports with the NRC.

Besides Stone & Webster's casual insouciance about safety, they happily colluded with Long Island Lighting Company, guaranteeing completion of the Shoreham plant within five years, but then moving the date back year after year while the government allowed it to charge up to half a billion a year for the "nearly finished" plant. Ten years later, in 1983,

> Stone & Webster were still finishing Shoreham, but the President of the Board was hat in hand, still wheedling more money from the government, swearing under oath that "the Shoreham Unit is complete as far as construction is concerned—except for the diesel engines," ready to load fuel in two months when he knew…they'd be lucky to complete construction within a year (*Vultures' Picnic,* 299).

Sued along with its partner LILCO, Stone & Webster was found by the court to have conspired with LILCO to defraud. The plant operated just one single day, but Stone & Webster made off with a billion dollars, "and despite the verdict, walked away with a $50,000 settlement" (*Vultures' Picnic,* 300).

Operating in Mississippi, Georgia, Alabama and Florida, **Southern** bought England's Southwest Electricity Company, even though there was a law (the U.S. Public Utility Holding Company Act) that made it illegal. No problem. Southern had their lobbyists get to work on it, and Congress changed the law. The unfortunate who blew the whistle on the deal died in a mysterious plane crash.

Southern's talent for creative accounting dates back to the time when they charged their customers for coal mined from their own mines, but loaded the freight cars with rocks instead; or the time when they charged their ratepayers $100 million for the use of "spare parts" which were never used, technically violating the Federal Energy Regulatory Commission's accounting regulations. No problem. Southern had Congress repeal the law, thus putting an end to the FERC's regulations.

Also lurking under NINAs re-branding, Reliant (Obama's favorite campaign bundler, formerly known as Houston Lighting and Power and shell cover for Brown & Root, notorious tentacle of Halliburton), together with CSW, owned two reactors known as the South Texas Project. Both companies got the state to order their customers to foot the bill to build the reactors by swearing hope to die they could complete the work in five years max at a cost of $1.2 billion. In real time, it took twelve years and $5.8 billion. Eventually both Reliant and its contractor, Halliburton's Brown & Root, were ruled "incompetent" by the Public Service Commission, and had to pony up more than a billion in fines and penalties. Still, they managed to unload several billions onto their customers, thanks to a little help from their friend, George W. Bush. Maybe that's why before their machinations put them into Chapter 11 they provisionally changed their name to NRG Corporation before changing it once again to NINA (*Vultures' Picnic*, 265).

Which brings us up to 2010–11, when, although the NRC once considered de-licensing Reliant for lack of "moral integrity," NRG and Southern won the contest for fresh cash—all $56 billion of it.

For the Obama administration, things like Shaw's history are water off a duck's back, but it looks like Shaw won't be building those two reactors in South Texas any time soon. Recently NRG and Toshiba decided to write off their initial $331 million investment in light of soaring financial risks created by the reactor disaster in Japan. NRG CEO, David Crane stated, "We have concluded that financially, this is the end of the line for us." But Shaw still entertains plans to construct the first new reactors for Southern Company, and new reactor plants in the UK as well! And guess who expects to get a 20% piece of the South Texas Project action? Tokyo Electric Power Company! (Remember how Fukushima got them a lot of press?) And Shaw is currently building a plant to turn plutonium from old worn out A-bombs into nuclear plant fuel (MOX). The NRC was so thrilled, it made sure to exempt Shaw from having to install any anti-terrorist security measures. The NRC commissioner who made sure to vote the exemption, Jeffrey Merrifield, now works for Shaw! And former Secretary of Energy, Spencer Abraham, who promoted the plan, is now Chairman of Areva USA, a partner in Shaw Areva MOX services (*Vultures' Picnic,* 310)! Shaw even scored the contract to clean up at Fukushima, all of which demonstrates that incompetence pays and— with government support— fraud pays even better, especially if you're in a hurry to turn the planet into one wall-to-wall exclusion zone.

113.
Where the Real News Happens...

On the 252nd day following our planetary catastrophe, between downpours, OccupySanFrancisco holds a weekly *tertulia*, an open discussion, in which all are invited to participate. Before the hour, a town crier (who just happens to be the facilitator) circles through the campground. "Mic check, mic check," she calls. "A discussion on the Rights of Mother Earth is taking place outside the library." The library is a designated tent with some 500 books and tapes, available to anyone who cares to borrow them.

The discussants are a disparate group: a fellow with an ecclesiastical background and a bicycle who cites the Church Fathers and their belief in a sacred earth; street people, many of them on the road, stopping here from distant cities; a young woman with iridescent peacock blue eye-glitter and a service dog; a highly articulate guy sporting dreads; and some on-lookers who drift in and out again. It's a self-moderating group, like the *tertulias*, which allowed Spanish political prisoners under Franco to survive with their spirits intact. Each speaker holds a talking stick, is allowed no more than five minutes, and before speaking a second time, all must make sure that everyone who might have wished to speak in the first round has had a chance.

The subject of today's discussion: the Declaration of Rights of Mother Earth, a document that forms part of the Constitution of the Nation of Bolivia. It's a free-flowing conversation that winds through the basic tenets of Eurocentric civilization, whose sacred texts accord Man dominion over all living things, but it also touches on the centrality of the very lowest forms of life—the archaea—without which no life is possible on earth or in the human digestive tract—a reference to the two participants who've brought their lunch over from the food tent. It includes a report by Peacock Glitter about the city of Portland, which floated a ballot measure to confer personhood on the Willamette River. An initiative proposed: to identify one spring, located either in the city of San Francisco or in the East Bay, and to develop a resolution to be brought before the city council to declare the spring as having the rights of personhood. Followed by a round of conversation exposing the double agendas of corporations: Monsanto under the guise of enhancing crop production to feed the earth's growing populations, actually sterilizing the earth's seed to gain control of the world's food supply; fracking companies (such as BP, Shell and Conoco) creating a dependence for a for-profit supply of uncontaminated water; Pfizer marketing arsenic-laced, carcinogenic chicken feed, insuring a steady supply of human cancer to medicate, and the nuclear industry that promises to rescue the world from global warming by producing supposedly carbon-emission-free energy while littering the planet with its radioactive waste.

It's a focused discussion that winds down with a vote to decide what aspect the group chooses to pursue in future meetings. I make my way toward the BART station, passing one of a number of TV sound trucks waiting nearby in case the tack squad comes calling. A man sits inside the cab. I knock on the glass, mime rolling the window down. The man opens the door.

"Are you a reporter?"

"Yes."

"You just missed an amazing discussion. Maybe you'd like to be out there where the real news is happening." I eye him quizzically. But he's wearing his suit and tie, the cab is warm, and outside it's begun to pour.

114.

Occupy Tokyo, and in the Provinces....

The Fukushima 100 are so mad they're not going to take it anymore. They've come from Fukushima, a hundred-strong to occupy the steps of the Ministry of Economics, Trade and Industry, Japan's nuclear industry administering body. The government won't evacuate their children, some 300 of whom are already showing signs of irradiated thyroids. It's been five months, but the government has yet to issue decontamination guidelines. The 100 swell to 200. They know the rap: there's serious contamination beyond the 12.5-mile exclusion zone. Hot spots have even been found as far away as Tokyo and Yokohama. They're occupying and they're not going away. They sit on the steps one day, two days, three days—as long as it takes to get the government to listen. On the third day they put out the call: 4,000 people join them. Women come from Iwate, from Date, from surrounding villages. Farmers come. They've lost their rice, they've lost their winter feed. They can't sell contaminated meat. They've lost their sole source of income to contamination. It doesn't help that on November 3 TEPCO admitted one of GE's doomed rectors was showing signs of nuclear fission, proof of fresh leakage. They read the *Mainichi Shimbun,* they read the *Asahi Shimbun.* They know that after the disaster, the government raised acceptable radiation levels for individuals from 1 to 20 millisieverts per year. They know that children—and women, too—are much more susceptible to radiation than men.

They know that in their meetings with officials, they've only gotten the runaround. All the officials want to talk about is cleaning up radioactive waste, never mind the children. One of the women, Rika Mashiko, speaks. Her husband has to continue working in Fukushima on his organic farm located 50 km from the reactors, but with her daughter, she has evacuated to a Tokyo suburb, where she works part-time to support them. Because she left voluntarily to protect her child, she receives no help from the government. The government only authorizes help for the 200,000 folks within the 12.5-mile exclusion zone—which is why they've kept the zone so unconscionably small. And these women know it. They want stronger protective measures against radiation. They want transparency and honesty from government. But Ayako Ooga says all the government wants to do is placate people like herself. She represents Fukushima Mothers Against Radiation. She wants assurances "that a similar accident will never happen again and that the government will protect us from radiation." Evidently they're not buying TEPCO's excuses that, but for the earthquake, but for the tsunami, every one would be OK. Word must have leaked out along with the radiation that those reactors were unsafe.

Hatsumi Ishimaru, a farmer from Genkai, has come to Tokyo to join the Fukushima 4,000. She's party to a lawsuit filed to prevent the faulty Genkai nuclear power station from starting up again. Today "women are at the head of the anti-nuclear campaign because we value life more than economic gain." They know better because after all, they're the ones to raise the children, they're the ones who make things grow. Ishimaru is not alone. Some forty people in Otsu on the Sea of Japan are suing for an injunction barring the two Japan Atomic Power Company's Tsuruga reactors from starting up again. In Otsu, they have plenty of reason. An accident at Tsuruga would contaminate Lake Biwa, from which the entire Kansai district gets its drinking and irrigation water. They know Tsuruga's two reactors sit on top of an earthquake fault, compounding the risk of catastrophic failure. They know Tsuruga Unit 1 is 40 years old and needs to be retired. They know its location on the Sea of Japan makes the station vulnerable to tsunamis. And the disaster at Fukushima has made it clear to them that government inspection standards fall pitifully short.

In some of the cities and villages, in the absence of government directives, regional and municipal governments have taken matters into their own hands. In August—after five months of inaction—the government issued decontamination targets. Said radiation expert Tomoya Yamauchi, "I really doubt their seriousness." He suggests current levels around Fukushima ought to be reduced 90%. Areas showing more than 20mSv per year are government-declared off limits. The government has set 50–60% reduction targets for areas showing 1 to 20mSv per year over 2 years, trusting to the impact of rain and wind and cesium-134's normal degradation to do the rest. Shunichi Tanaka, former chairman of the Atomic Energy Society of Japan, says the government's goal of using local efforts to reduce contamination by 10 to 20% is woefully inadequate. The government's use of high-pressure washers is not enough. Yamauchi says drastic, high-cost efforts are needed to replace roofs and the surface of asphalt roads, and so far there is no talk of decontaminating the 20-km exclusion zone. Tanaka stresses that because the decontamination effort involves the people actually living in affected areas, their participation in the process is essential. He directs cleanup efforts at Fukushima, and advises the municipal government of Date, a highly contaminated adjacent town. Since there is no government-mandated plan yet for storage of contaminated waste, local residents need to be involved in finding temporary space.

115.
Unwanted Earth

What do you do with a heap of radioactive waste and soil? No problem. Rather than wait for the government to carry out decontamination efforts, the city of Fukushima has issued a citizen's Q & A do-it-yourself manual:

> Q: What's the first basic step?
> Q: Should disposable protective outfits be worn for decontamination work?
> Q: Where should one start the decontamination?
> Q: Where are hotspots likely to show up other than at the bottom of downspouts?
> Q: What should you do with the soil and leaves?

Q: What should be done after completing the work?
Q: Where will tainted water used in decontamination work eventually end up?
Q: Will decontamination reduce radiation levels to pre-March 11 levels?
Q: Do people in Tokyo also need to decontaminate their property?

The City of Fukushima plans to place its priorities on cleaning up places where people carry out their daily lives: schools, school routes, houses, stores and public facilities. The City of Date has set up a one-stop decontamination facility with a foodstuff radiation testing service where townsfolk can get cleaning tools, high pressure washers, and storage bags. TV programs alert people that help is available free of charge. It may go without notice that the citizens who take up these offers, determined to "do-it-yourself" are needlessly and tragically exposing themselves to the hazards of radiation.

If you thought do-it-yourself decontamination might be fun, think again. Which would you prefer: climbing on your roof disguised in hat, face mask, gloves, waterproof boots, and disposable pants and long sleeved shirts to take a scrub brush and a solution of vinegar and bicarbonate of soda, or laundry starch to scour no-see-ums off your roof tiles, then digging up 5 cm of top soil all over your yard and bagging it, along with all your weeds and grass and vegetation—or installing solar panels on your roof?

Two years ago, the City of Yokohama may have asked such a question. Two years later, 5,000 solar panels have been installed on 5,000 roofs with 4,000 more to go in the next three years, making all Yokohama homes micro-sized power companies, capable of generating all they need and selling surplus back to TEPCO's power grid. It's cleaner and less bother than getting up on your roof in gumboots and spraying laundry starch in all the nooks and crannies between each of your roofing tiles.

Meanwhile, the Tokyo government continues doing what comes naturally. Forget the gumboots, pressure washers and liquid starch. Like all governments, what Tokyo does best is talk. The Government Revitalization Unit kicked off a four-day-long *shiwake*, a panel talk session roughly translating as an "investigation," bringing together those inevitable government lawyers and predictable private-sector energy experts under new PM Yoshihiko Noda's watchful, pro-nuclear eyes. Their recommendations: radically revising (but not altogether quashing) plans for the troubled Monju prototype fast-breeder reactor; canceling a 2.2 billion yen allocation for Monju in 2012; and zeroing out spending on ITER, the experimental fusion reactor project located in Southern France in which Japan forms part of a consortium along with China, Russia, South Korea, India and the EU.

But for all the talk, the Fukushima 100, 200, 4,000 are not going away. In the heart of Tokyo, they occupy the steps of the Ministry of Economics, Trade and Industry, demanding action. They want the government to evacuate their children from the contaminated zones, zones that, according to the *Asahi Shimbun* of November 21 cover one-eighth of Honshu or 30,000 square kilometers.

116.
Dream of the Fukushima One Hundred

They're propping the doors open, evacuating the Ministry. It must be an emergency. Rescue workers, white hard hats, blue uniforms tucked into their gumboots. One or two of them at first. Running down the steps. Pulling something. Some kind of hose, yellow, thick as sewer pipe, bumping it down the steps. It must be heavy, so many of them, really straining, pulling on it hard, dragging it along the sidewalk. And there's Yoshihiko Noda! The prime minister, just like in his photographs! And Goshi Hosono, the disaster czar, what's he doing in a hard hat? And Prime Minister Kan—just like his picture in the paper! They must be part of the disaster, hauling the pressure hose, snaking it like a long pipe all along the sidewalk, and all the representatives lending them a hand. And red-eyed Yukio Edano, there he is, bringing up the rear, dragging the tail end along the ground, stretching it like a long yellow caterpillar, winding over the hills, over the roads that run beside the sea, past all the broken up streets and splintered houses, hauling it out to Fukushima because the trains don't run anymore. And buckets and big scrub brushes to power wash the roofs and hose them down with vinegar.

Morning already! All the women—and some men—farmers like us, everyone going down the long, dark corridor, and inside the heavy doors where the dark velvet benches are all lined up in a circle. Everyone sitting on them—and some men too—farmers like us. Talking, saying the same things, back and forth, calling on each other, saying the things we have to say. Talking into the microphone because we can't talk very loud, not loud enough to hear. And we're talking, and it's noon already and we go on talking. . . . Getting ready.

117.
Wrong Way on a One-Way Street

What if James Watt had been born under French skies, and not in Scotland? Or what if Augustin Mouchot's invention of the solar-powered steam engine had preceded (rather than succeeded) Watt's by 100 years? The history of industrial age energy might have taken a radically different course. As it is, from its beginning its Anglocentric history projected a trajectory similar to traveling on the wrong side of a one-way street. Instead of looking below the Earth's crust for ever-declining resources to fuel the industrial age's growing appetite for energy, it might have been wiser to look above to the sun, as Mouchot did, for the ever-renewable energy it could provide in limitless quantities.

Day 266 following the planetary disaster at Fukushima, the morning dawns shrouded in fog so dense, the tall redwoods lining the opposite side of the street hover like ghosts, their crowns barely discernible at 40 feet. It reminds me of what Londoners called pea-soup fog, the result of burning coal in all London's fireplaces against the winter chill.

How clear were the Scottish skies when James Watt made his appearance on earth in 1736? How clear were the skies in 1825 when

Augustin Mouchot was born? Why did one man look beneath the earth to power his engine, while another looked to the sun? Long before the year of Watt's birth, the so-called "primitives" of the Southwest had thought to orient their winter cliff dwellings on the north sides of their steep canyon homes to benefit during the colder months from the sun's radiation. Even proponents of *Feng Shui* will tell you a front door opening east is always propitious. How did it occur to a European to see with eco-centric eyes and with a sensibility that concerned itself with implications for the future?

Watt and Mouchot were very different men in terms of social class, culture, and sensibility. While Watt was born into the merchant class, lacking skill for any scholarship other than mathematics and mechanics, Mouchot was born in the Morvan, a district in the heart of France known for its natural beauty. Except for the year in which the Ministry of Education gave him a grant and a leave of absence to travel to Algeria to perfect his invention of a solar steam engine, he served as a schoolmaster, moving from teaching grammar school in the Morvan to teaching high school in Tours and later in Rennes.

> One must not believe, despite the silence of modern writings, that the idea of using solar heat for mechanical operations is recent. On the contrary, one must recognize that this idea is very ancient and its slow development across the centuries has given birth to various curious devices.

What curious devices might Mouchot have referred to in his 1880 statement? *Chaleur Solaire,* the small book he published in 1869, the same year he displayed one of his first models, credits the Arabs with their glass-making skills for experiments focusing the sun's rays to obtain heat, and traces their interest to ancient Egypt and later Greece, evidence of Mouchot's historic sensibility. Even more telling, in another short section he points out the sun's role in regulating the planet's wind and ocean currents, and nurturing the life of plants and animals, suggesting a deeply ecological cast of mind. But, unlike Watt, he never succeeded in partnering with anyone to obtain support for his invention. He taught school until his retirement. One can only imagine how fortunate were those pupils who studied under him.

By 1869, the year he displayed one of his first prototypes, the price of coal had dropped, effectively making his invention all but irrelevant in the eyes of the public. "Eventually industry will no longer find in Europe the resources to satisfy its prodigious expansion.... Coal will undoubtedly be used up. What will industry do then?" Mouchot wrote those words in 1878, the year he exhibited, at the Universal Exposition in Paris, the great axicon he had perfected—essentially a solar dish. The earlier model, which he displayed in 1869, disappeared in 1871 during the chaos and destruction of the Franco–Prussian War, a war in which, with the German annexation of the mining district of Alsace-Lorraine, France lost its access to cheap coal—which may in part explain why present-day France is the European country relying most on nuclear energy.

It's hard to imagine what the consequences to our planet might have been, had Mouchot's timing been otherwise, had the Industrial Revolution not resulted in the growing infrastructure that guaranteed support for Watt's invention to the exclusion of Mouchot's: the coal mines in which children and women worked sixteen hours a day stripped to the waist to better withstand the heat, crawling on their knees to push the coal cars in galleries too shallow to allow for mules or machines to do the work; how thousands died of tuberculosis in the crowded cities of newly industrialized Europe whose skies became darkened by carbon pollution, with air so thick, at dusk you could see the carbon particles shimmer in the fading light.

118.
Fortune Cookies

"We can't respond to every seismologist's claim," *especially* if it runs counter to a formidable pile-up of vested interests, firms like Panasonic, Kawasaki Heavy Industries, Ltd., and Komatsu, Ltd.

Bureaucrats like Masao Urakami, Japan Atomic Power Co. spokesman, are not known to speak in full paragraphs. They prefer truncated messages that stop before "especially." Seismologist Katsuhiko Ishibashi, whose career is studded with truncated fortune-cookie-type communications from bureaucrats, knows this all to well.

Ishibashi started student life in Tokyo, where the 7.5 quake that killed dozens in Niigata shook him up so much that he decided to specialize in seismology, which until that time had focused on a history of written records to the exclusion of actual geological investigation in the field. But throughout his career, his prognostications have met with denials at every turn.

In 1994, Ishibashi's boss made him apologize publicly to the Construction Ministry for publishing his claim that Japan's existing building codes put cities at risk. Five months later, when thousands died in the Kobe earthquake, *A Seismologist Warns* made the bestseller list. In 1997, although he was by then a professor at Kobe University, when he started naming the deadly cocktail of earthquake risk combined with nuclear energy "nuclear earthquake disaster," he was blown off as a "rank amateur" by Haruki Madarame, who now heads Japan's Nuclear Safety Commission. "In the field of nuclear engineering," Madarame stated, "Ishibashi is a nobody." But last month Madarame had to eat crow when he conceded that "because of the [Fukushima] accident, there's a need to take another look at things....Ishibashi contributed a lot to the revisions of the earthquake guidelines...." For his part, Madarame contributed, at least to some degree, to 200,000 people losing their homes, their land, and their way of life, and, according to one of the Fukushima 100, "tearing our families apart." But perhaps Madarame's list of nobodies is long enough to add at least another 200,000 names.

Ishibashi advocates ranking Japan's plants by risk and phasing out the worst among them. "What was missing and is still missing is a recognition of the danger." Japan Atomic Power Co., the operator of the Tsuruga Plant, recognized only four years ago that an active fault ran under it. Even though JAPCO believes Tsuruga will be safe after reinforcing it, 40 of the Fukushima 100 are suing to shut it down. Referring to Ishibashi, JAPCO's spokes-bureaucrat Masao Urakami opines, "we can't respond to every claim by every scientist. Standards for seismic ground motion are based on findings by experts assigned by the government." Presumably one of the experts he has in mind is Madarame, although it's conceivable that in the larger scheme, Mother Earth may be the "decider," notwithstanding her failure to be recognized by the Japanese government—or any other government for that matter—with the exception of the government of Bolivia.

Although Ishibashi must already have had his fill of butting heads against all the bureaucratic 'yes, buts,' he keeps running up against a convoluted repertoire of excuses about why shutting down is not an option. It can't be done because:

• without nuclear energy, Japan's industrial sector will have to pony up an additional 1.7 trillion yen ($22 billion) to cover its yearly energy bill. (It prefers paying now to paying later for disaster cleanup. In any case, in a recent lawsuit against TEPCO, TEPCO won by pointing out that scattered radioactive particles no longer belonged to them (Jason Clenfield. "Vindicated Siesmologist Says Japan Still Underestimates Threat to Reactors").
• industries like Komatsu may threaten to move overseas if a stable supply of electricity can't be guaranteed.
• power companies need to underestimate risks to minimize construction costs. "If we engineered factoring in every possible worst case scenario, nothing would get built," says Masamori Hamada, member of a government panel revising Japan's nuclear plant standards. "What engineers look for is consensus from the seismologists and we don't get that."

"Nuclear plants are such a huge investment that operators need some assurance of getting their money back," says Norm Abramson, a UC Berkeley seismologist, "but regulatory stability and scientific change don't go hand in hand."

Minutes of a 2009 Trade Ministry meeting on safety at Fukushima Daiichi show that TEPCO and the government regulator pushed aside scientific findings that emerged only after the power station went on line. Isao Nishimura dismissed the omission of studies showing that northeastern Japan had a history of major earthquakes and tsunamis by providing his own fortune cookie reading: "We didn't think the damage would be that significant." He left the part out that says "and we didn't care." Debate at that meeting was cut short by a Nuclear and Industrial Safety Agency official, and despite studies by another seismologist, Koji Minamura, showing that giant tsunamis had smashed into Japan's northeast coast every thousand years or so, the regulator approved Fukushima Daiichi's safety report barely a month later.

Aborting meetings seems to be a favored nuclear industry strategy. Before waiters began to serve dinner at a recent meeting of the Kobe Chamber of Commerce held in a Chinese restaurant, Ishibashi was able to get through only a few of his 36 power-point slides warning of the risk of having thirteen more-than-30-years-old nuclear reactors in a highly active seismic area—reactors that keep the Panasonic and Kawasaki factories running and power up the cities of Osaka, Kyoto and Kobe, before waiters began serving dinner. At the end of the evening, the CEO of one of Japan's biggest companies leaned over the fortune cookies to whisper to Ishibashi, "I know you want the reactors shut, but it can't happen. We need the electricity" (Bloomberg News. "Vindicated Seismologist Says Japan Still Underestimates Threat to Reactors").

119.

And in the U.S. a Nuclear Safety Check List…

Gregg Levine writing for *Capitoilette* on Dec. 9 the 273rd day after Fukushima provides an update on Davis-Besse—and all the cracks overall—appearing in the U.S. nuclear industry's safety mechanisms:

> Either safety—with regards to protocols, equipment and people—is up to snuff, or it is not. As Jaczko [director of the NRC] observes—and the many "unusual events" he has had to deal with this year make clear—the safety of America's nuclear reactors is not where it needs to be. The accidents at Cooper Nuclear Station in Nebraska and the Perry Nuclear Power Plant in Ohio most definitively exposed workers to high (and possibly dangerously high) levels of radiation….
>
> [Other] plants having significant issues—Crystal River in Florida, where news of a third major crack in the containment building recently came to light…, Nebraska's

Fort Calhoun, which is still shut down after flooding earlier this year...; New Hampshire's Seabrook, where crumbling concrete was discovered in November, a month after the plant had to shut down because of low water levels...; Vermont Yankee, where radioactive tritium continues to leak into the Connecticut River...; North Anna, which... scrammed when the Mineral Springs, VA, earthquake shook the reactors well in excess of their designed tolerances...; Calvert Cliffs in Maryland, where a piece of siding blown off by Hurricane Irene shorted a transformer, and the resulting loss of power to safety systems caused its reactor to scram...; Michigan's Palisades nuclear power plant, which had to vent radioactive steam when it scrammed after worker error triggered a series of electrical issues....

[Or] constantly troubled Davis-Besse plant near Toledo, Ohio, [where] the Commission just (as in 4:40 PM on Friday, December 2) okayed a restart there, despite serious concerns about numerous cracks in its shield building. (On December 7, one day after the reactor restart, FirstEnergy, Davis-Besse's operator, admitted that they had withheld news of new cracks on a different part of the structure, which were discovered in November.., [but] they insisted they only withheld the information from the public, and that they did report it to regulators—which raises grave questions about the honesty, independence and competency of the NRC and how it could approve a restart.)

Given the evidence—and given that the NRC itself spent all summer studying the crisis and drafting recommendations based on "lessons learned," it is hard to believe complacency is really the problem. It is probably even too generous to say that the industry suffers from willful ignorance. No, when considering the contagion spreading from Japan and the coughs and hiccups here in the United States, it is probably more accurate to say that the profit-driven, government-protected nuclear sector is actively callous (Gregg Levine. "Gregory Jaczko Has a Cold.").

120.

Mission Accomplished...in 30 to 40 Years

On Day 280, PM Yoshihiko Noda declared an end to the nuclear crisis, stating that technicians have regained control of the reactors at Fukushima Daiichi nuclear power plant. "Today we have reached a Great Milestone," he trumpeted in a December 16 TV address to the nation. "The reactors are stable, which should resolve one big cause of concern for us all."

In keeping with his conservative taste, Mr. Noda left his carrier flight deck outfit at home, preferring the simple 3-piece camouflage favored by Japanese functionaries everywhere. Footnotes were not supplied clarifying which reactors he was referring to, what if any other causes of concern there might be about which information was being (temporarily) withheld, or the population segment he might be referring to as "us" (Hiroko Tabuchi. "Japan's Prime Minister Declares Fukushima Plant Stable").

Mission accomplished notwithstanding, Mr. Noda, 54, emphasized his commitment to "fight to the end," acknowledging that dismantling the plant will take anywhere from 30 to 40 years. Mr. Noda did not say he might be 94 when the fight concludes.

121.

Ten Months, Ten Days...

As of Day 283, only six of Japan's nuclear plants are currently shut down, most for stress test evaluation. Braving the winter cold, women have intensified their vigil in front of the Ministry of Economic Affairs by setting up tents, preparing for a sit-in campaign to last ten months and ten days, the time traditionally acknowledged in Japan as the length of a full-term pregnancy.

"Our protests are aimed at achieving a rebirth in Japanese society," said Chieko Shiina, a Fukushima grandmother. "There is a need to change the way the authorities have run the country by putting economic growth ahead of protecting the lives of people." "We do not trust the government anymore," added Takanobu Kobayashi, who heads the Matsuda network of citizen's movements.

The meltdowns at the crippled Fukushima Daiichi nuclear plant have prompted an unprecedented rise of Japanese women leading anti-nuclear protests. "Mothers are at the forefront of various grassroots movements, working together to stop the operation of all nuclear plants in Japan by 2012," said Aileen Miyoko Smith, head of Green Action, an NGO promoting renewable energy.

Day 324, Beyond Nuclear issues a report that the tent occupation of the Ministry of Economics, Trade and Industry by the peacefully protesting mothers of Fukushima has been threatened with eviction by order of Minister Yukio Edano, who has presumably bowed to pressure by the nuclear industry following a "public" meeting where concerned citizens who registered were forcibly removed, allowing the meeting to continue behind closed doors (Beyond Nuclear. "Fukushima Mothers and Activists face eviction").

Protest movements have not garnered much support in the past in a society valuing conformity and hard work, and protests led by women especially are unprecedented. Now such demonstrations have spread to cities throughout Japan (S. Kakuchi. "Mothers Rise Against Nuclear Power," *Common Dreams*). On Day 330, 51 of Japan's 54 reactors are currently shut down for "routine maintenance."

122.

Meanwhile at the Ranch...

Tomohiko Suzuki, a freelance journalist, shares his findings on Day 279 at the Foreign Correspondence Club of Japan. Working five weeks undercover as a general laborer at Fukushima Daiichi, he outlines how TEPCO is showing a marked concern for appearances over the safety of employees or the public.

> Conditions are far worse than [TEPCO] or the government has admitted. Much of the work is simply for show and fraught with corporate competitiveness. Reactor makers Toshiba and Hitachi, brought in to resolve the crisis, each have their own technology and they don't talk to each other....

Suzuki claims the "cold shutdown" schedule has promoted a disregard for worker safety. Workers are assigned projects that can't possibly be completed in the time allotted without manipulating dosage numbers. Radiation screenings are essentially theater. The radiation detector is passed too quickly over each worker, and the line to the warning buzzer has been cut. Suzuki quoted a nuclear-related company source saying that working at Fukushima is equivalent to a "death sentence." Suzuki, who has authored a book, *Yakuza to Genpatsu,* (trans. *The Yakuza and Nuclear Power*) claims that people unable to repay loans made by yakuza gangs have been forced to work at the site. He alleges one in ten laborers is aligned with criminal organizations in one way or another.

A breakneck work schedule, leading to shoddy repairs (including the use of plastic piping likely to freeze and crack in winter), has choked off any new ideas. Hitachi and Toshiba engineers presenting new solutions have simply been told there isn't any money to try them out ("Reporter Works at Fukushima, Exposes Real Situation at the Plant," *SimplyInfo*, December 17, 2011).

123.

Waiting for the Other Shoe...

At the time of the earthquake of March 11, which destroyed all six GE Mark I reactors, Unit 4 was being refueled. Refueling procedure requires that core assemblies be raised to the level, five stories above ground, where the stored fuel pool is located. Reports began to surface as early as December 5 that Unit 4 had lost its south side wall. On December 2, beside Reactor 4 something like "fire" was observed on camera. Since then, a strong light has been set toward the

Fukuichi webcam as if to hide something by white out. But what? In an interview with Thom Hartmann, Paul Gunter, director of Beyond Nuclear's Reactor Oversight Project, stated, "Unit 4 is looking more and more like the leaning Tower of Pisa....The whole building is structurally listing. The full core was offloaded to the [cooling] pool during refueling, and that building is now shifting. [TEPCO] has been sending engineers in there to try and shore it up with poles and whatever to keep it from falling over" (WACEB, "Nuclear Expert: 'Unit 4 is looking more and more like the Leaning Tower of Pisa right now'").

"If the fuel rods spill onto the ground, disaster will ensue and contaminate Tokyo and Yokohama, creating a gigantic evacuation zone," wrote Akio Matsumura in September of 2011.

> All scientists I talked with say that if the structure collapses, we will be in a situation well beyond where science has ever gone. The destiny of Japan will be changed, and the disaster will certainly compromise the security of neighboring countries and the rest of the world in terms of health, migration, and geopolitics (Akio Matsumura, "The Fourth Reactor and the Destiny of Japan," September 29, 2011).

Simply stated, according to Matsumura, Tokyo and Yokohama will become uninhabitable zones for all eternity.

124.
Fukushima You

A new epidemiological study by Mangano and Sherman suggests that in the United States in the 14 weeks following the Fukushima meltdown, there were 14,000 excess deaths ("An Unexpected Mortality Increase In The United States Follows Arrival Of The Radioactive Plume From Fukushima: Is There A Correlation?" *International Journal of Health Services,* Volume 42, Number 1, 2012). This study follows Mangano and Sherman's finding that infant mortality in 8 northwestern United States cities rose 35% in the three-month period following March 11 (Janette D. Sherman, M.D. and Joseph Mangano:

"Is the Dramatic Increase in Baby Deaths in the U.S. a result of Fukushima Fallout?" Counterpunch, June 12, 2011).

None of these considerations seem to have carried much weight with the Nuclear Regulatory Commission. December 22, not four days after Gregory Jaczko returned from his inspection tour of Fukushima Units 1 and 2, and Day 290 following the Fukushima planetary disaster, the NRC announced it had unanimously approved Southern Company's radical new nuclear plant design for the Westinghouse AP1000, a 1,154 megawatt reactor with an "advanced passive design." The commission even waived aside the 30-day waiting period for the approval to become official. Jaczko claimed that all of the panel's safety concerns had been fully addressed.

Together with South Carolina Power and their partners, Southern Company (see Section 112, "NRG and Southern: Evil Twins") doesn't seem to worry construction approval won't be forthcoming for their four new reactors, two planned for a location near Augusta, Georgia, and two others at the Summer plant in South Carolina. They have already spent hundreds of millions of dollars digging foundations for the projects and taken other steps that don't require NRC approval ahead of time (*New York Times,* December 22, 2011).

125.
Happy Talk, or, Why Gregory Jaczko Was Hanging Out in Tokyo

If NRC chief, Gregory Jaczko bears an uncanny resemblance to Humpty Dumpty, the similarity if unfortunate, is perhaps not inappropriate. On day 288, two days following his December 18 official tour of Fukushima Units 1 and 2, Mr. Jaczko held a news conference at the United States Embassy. Commenting on TEPCO's efforts to bring the nuclear crisis under control with "cold shut down," Mr. Jaczko happily declared that there was no longer enough energy in the crippled Fukushima Unit 1 nuclear reactor to produce an offsite release of radiation. Presumably his tour excluded a visit to Unit 4 ("U.S. Nuclear Chief says Fukushima plant stable but major task remains," *Mainichi Daily News*, January, 19, 2012).

Mr. Hosono, Japan's Minister in Charge of Nuclear Affairs, followed on Mr. Jaczko's coattails. "Thanks to support from the U.S., the onsite accident has been brought under control," he breathed. "If we [get] assistance from the U.S. in the decommissioning of the reactors, which will take another 30 years or more, we will certainly overcome this to make Fukushima safe again. Government officials...pledge to enforce safety regulations strictly and to insure transparency."

126.
Day 274. Behind Closed Doors

I wake from the sound of a woman gagging. No way to describe such a sound: an explosion from the chest, filtered through a wall of tears. An auditory dream. No image, just the sound of a woman choking as if something is being forced down her throat.

I know this sound. I recognize it. It is a sound without an image. I am in a hospital corridor. Behind the door, I can hear my mother screaming. She is 91 years old. She is a stroke victim. Behind the door where I can't see, they are forcing a gastric tube down her throat. She is screaming, wailing, her throat is being raped against her will. Now they pour milk down the tube. My mother doesn't digest milk. She will soil herself until the bed is drowning in her own waste. But she is without speech. The stroke has stolen all her words.

After, I stand at her bedside. One blink for yes, two blinks for no.

Is the discomfort here, in your head?
Two blinks.
Is the discomfort in your neck?
Two blinks.
In your chest?
Two blinks.
In your arms or hands?
Two blinks.
In your lower body?
One blink.
I raise the sheet.

The wetlands ooze with it, the sticky red. Birds are coated in it, and sea turtles. When they swim to the surface, they die asphyxiated. The cleanup crews are denied protective masks. It looks bad on camera. They're black people anyway.

Mother Earth is screaming. She is in a room behind closed doors. They are forcing her to drink: strontium, cesium, plutonium, americium, cobalt, fracking compounds, Corexit®.

I burst in the door that time. Stop it. Stop the tube. Stop choking her. She has a directive. The healthcare directive of Mother Earth. You can't waterboard your mother anymore.

In Durban, South Africa, where the last day of climate talks are taking place, Mohamed Aslam, Environment Minister of the Maldives, one of the lowest lying islands in the world, speaks: "You need to save us. The islands can't sink. We have our rights. We have a right to live. We have a right for home. You can't decide our destiny!"

But the U.S. climate negotiators have their marching orders. No, not right now, they say. Maybe in five years, or maybe ten. The Obama Administration will "cook Africa" where temperatures are expected to rise to 150 degrees, and disregard the 48 extreme weather emergencies declared this year in continental United States, where the criminal indifference of earth-destroying corporations is aided and abetted by government complicity.

On the last day of the conference, the youth representative, a US college student, speaks. Anjali Appadurai's hair is long and black and wild. Unlike the other representatives, she does not wear a suit:

> "I speak for more than half the world's population. We are the silent majority. You've given us a seat in this hall, but our interests are not on the table. What does it take to get a stake in this game? Lobbyists? Corporate influence? Money? You've been negotiating all my life. In that time, you've failed to meet pledges, you've missed targets, and you've broken promises....
>
> We're in Africa, home to communities on the front line of climate change. The world's poorest countries need funding for adaptation now. The Horn of Africa and those nearby in KwaMashu needed it yesterday. But as 2012 dawns,

> our Green Climate Fund remains empty. The International Energy Agency tells us we have five years until the window to avoid irreversible climate change closes. The science tells us that we have five years maximum. [Yet] you're saying, "Give us 10."
>
> The most stark betrayal of your generation's responsibility to ours is that you call this "ambition." There is real ambition in this room, but it's been dismissed as radical. Long-term thinking is not radical. What's radical is to completely alter the planet's climate, to betray the future of my generation, and to condemn millions to death by climate change.
>
> In the long run, these will be seen as the defining moments of an era in which narrow self-interest prevailed over science, reason and common compassion.
>
> Her time is up. The moderator speaks: "Thank you. Miss Appadurai...was speaking on behalf of half of the world's population, I think she said at the beginning. And on a purely personal note, I wonder why we let half the world's population speak not first but only last."

And in Durban last night, a climate-collapse-related extreme weather event: a storm of such severity that it killed eight people whose houses fell on them. And today Russian scientists who've been monitoring the release of methane in the arctic ocean publish their findings: "Earlier we found torch-like structures only tens of meters in diameter, but now we are finding continuous, powerful and impressive seeping structures more than 1,000 meters in diameter" (Derrick O'Keefe. "After Durban: We Must Pull the Emergency Brake Before the 1 Per Cent Drive Us Off the Cliff").

127.
Day 275. Behind the Lines

Dr. Hermann Scheer, who died in October, 2010, was a pioneer in renewable energy and director of EUROSOLAR, the European

Association for Renewable Energy, and Chairman of the World Council for Renewable Energy. If we take anything from Hermann Scheer's clotted prose, it's the understanding that change, real change, radical change, is fated never to happen in the halls of power, not in the halls of Durban, or Doha, or Cancun, or Copenhagen, or Seattle. Real change happens from the bottom up, from places where people wear their hair wild, or in rococo city council rooms where small municipalities practice the *quand même* of their own lights, where small incremental initiatives are born to the cheering of the people who have small stakes, who own small houses or storefront businesses, on modest streets, not from the Mighty who practice choking on their words.

> The strategist looking for a breakthrough to renewable energy should… direct her or his attention towards three points:
> •towards energy availability that is widely dispersed and independent, instead of concentration on particularly 'economical' international sites;
> •towards political decentralization, instead of towards international institutions and "market harmonization";
> •towards stimulating autonomous investments, instead of towards investment planning by government and the energy business (Scheer. *Energy Autonomy*, 239).

Scheer suggests that renewable energy will increasingly catch on to the degree that it is not tied to the grid. And he advances the idea that financing has much to learn from the microcredit movement. Wherever possible, he advocates legislation that fixes price regulation, and a tax structure that stops privileging the nuclear industry and eliminates tax breaks for aircraft and shipping fuel. In light of the failure of any followup to a proposal that the World Bank expand its energy portfolio to concentrate on a shift to renewable energy, he proposes establishing an International Bank for Renewable Energy and Energy Efficiency such as the one Michael Eckart, president of the American Council on Renewable Energy, proposed as part of his Solar Bank program.

Scheer turns much of his attention to education that will dispel sociological and politically-based resistances to moving away from nuclear energy, and emphasizing renewables. He identifies four of renewable energy's advantages over fossil and nuclear fuels:

> •The use of nuclear and fossil energy entails serious environmental degradation, with tectonic consequences starting with initial production, continuing until their byproducts are emitted into water, air, and the atmosphere. Renewables in general are free of such consequences.
> •In contrast to renewables, fossil energy can be depleted, resulting in rising costs and supply bottlenecks and emergences.
> •Nuclear and fossil energy reserves lie in relatively limited areas around the globe, resulting in lengthy supply chains to facilitate their use and major infrastructural outlays leading to growing dependence and provoking economic, political and military conflicts. By contrast, renewables can potentially result in macroeconomic efficiency, political independence and peacekeeping.
> •Because their reserves are scattered throughout the world, fossil and nuclear energy are becoming increasingly expensive, both in terms of direct and indirect costs, whereas renewables, because their cost becomes increasingly cheaper with technological improvements, mass production, and ingenious new forms of application, have the potential to result in enhanced social welfare and better economic strategy.

Can renewable energies sustain the lifestyle that industrialized countries have come to take for granted? That is the big question. The response is a complex one; the change will be a qualitative one, primarily based on the need for de-centralization. However, from the quantitative perspective, Scheer has this to say:

> the sun and its derivative (wind, waves, water and biomass) 'deliver' a daily dose of energy that is 15,000 times greater

> than what we now consume in the form of nuclear and fossil energy. To speak of insufficient energy potential is therefore downright laughable (*Energy Autonomy,* 48).

With the signing of the PURPA act by Jimmy Carter in 1978 mandating centralized U.S. energy systems to allow other suppliers to use the energy grid, local capacity resources entered the energy mix. Taken together, such preferred resources as energy efficiency, demand-response, distributed generation from a wide variety of renewables, combined heat and power (often referred to as co-generation) and improving storage technologies, all add up to significant savings when compared to fossil-fuel-derived energy.

Taken one by one, such elements may lack "elegance," let alone "splash," but together they undercut the price of natural gas by a significant amount, at the same time reducing the economic motive driving fracking. Even something as obvious-but-unsexy as planting trees in the city of Sacramento has been demonstrated to reduce peak temperatures by seven percent.

According to WEM's Barbara George, there is no reasonable price of solar PV, at any scale, including residential, that would not undercut the cost associated with a new natural gas peaker plant. From 2012 to 2020, peak summer energy demand is projected to remain fairly constant at 43,000 megawatts, a period during which California's excess energy averages 24,000 megawatts, some 3,000 supplied by nuclear power.

Taken together, such developments have the potential not only of slowing the process of climate collapse but effectively obviating the need for any form of nuclear-generated energy.

Although Japan partially deregulated its energy industry 16 years ago, the 10 Japanese regional power utilities still control 98 percent of the electricity market, but an emerging block of clean energy supporters is now beginning to see a share of the country's 15 trillion yen market. With 70% of the Japanese public now opposed to nuclear power, the government is encouraging conservation programs, and public figures have started to take the lead, with Masayoshi Son, CEO of Softbank Corporation (and Japan's richest man—take note, Warren Buffet) proposing a "supergrid" that could efficiently transmit energy across the country, allowing renewable energy to meet 60% of Japan's needs by 2030.

Some two years ago, the city of Yokohama launched a modest experiment designed to save power and cut carbon emissions: it encouraged its citizens to install solar panels on their roofs and to buy equipment to track how much energy they use. In the light of the nuclear disaster at Fukushima, what started out as a small-scale environmental plan has become a model development that has turned Yokohama households into micro-sized power companies able to generate their own electricity and sell the surplus back to TEPCO's grid. However, to access the grid, they still have to pay a tax of 15 to 25 percent. A rather timid law, which takes effect in July, requires utilities to purchase a certain amount of energy from renewable sources but at a yet undecided price. But the Yokohama model effectively turns the utility into a bystander rather than a buyer. City officials say that within three years a total of 9,000 homes will be panel-equipped, up from 5,000 homes today. Even so, Yokohama is still at TEPCO's mercy because it needs to negotiate rental fees for the use of an electrical storage facility and to pry loose prior use records assessing energy savings—which TEPCO is not about to part with, possibly because TEPCO never kept any.

128.
Day 281. It Starts With One

Budgets have never worked for me. I don't have the mentality, and I don't want to spare the time. It helps me to think in terms of just making do with less, or as little as possible. It's what environmentalists have come to call powering down. Federal agencies and governments in the industrialized world have demonstrated at Durban their intention to continue dragging their feet while continents cook, and extreme weather events dry, swamp, burn, shake, starve and blow the planet away, requiring people to think locally more than ever before. In his excellent compendium published this year by the International Forum on Globalization, Gar Smith puts it like this:

> To minimize cascading climate casualties and impending wars over shrinking resources, industrialized economies need to formally recognize the limits of Earth's carrying capacities, put the brakes on growth-oriented economics, and begin a wide-ranging process of...learning to live better with less, to unplug from the grid, to simplify our lives, to become more localized and self-reliant. The watchwords...'less and local,' 'carbon footprint,' 'slow food,' 'locavore'...have come to mean a change in our values and habits—a rejection of economies based on consumption and growth in favor of solutions that emphasize localization and less use of energy and materials in all economic activity"—not to mention total closed-loop recycling. Smith has an energy road map to propose. It begins with a unit of one:
>
> 1. Joining the "the 2,000 Watt Society." If the average Swiss citizen can manage to live comfortably on 5,000 watts (17,520 kWh per year), Americans, who presently consume 12,000, can conceivably learn, with mindfulness and conservation, to power down to, say, 6,000 for starters.
>
> 2. Zero Energy buildings. Learning from billion-year-old creatures who figured out sustainable technology to rethink building design. The ideal building uses only the energy it generates. Such a building exists in Harare, Zimbabwe. Designed by Zimbabwean architect Michael Pearse, the Eastgate Office Complex makes use of ventilation principles learned from the termite mounds of Zimbabwe. It uses 90% less energy than similarly sized buildings, saving of more than 3.5 million in construction costs by not installing air conditioning.
>
> 3. Starting One Block at a Time. Communities are forming aggregates to purchase discounted-rate power. Some cities

operate their own power plants. 75% of Aspen's electricity is generated from wind and other renewables. One Block at a Time brings neighbors together block by block to lower solarization costs.

4. The Transition Town Movement is gathering steam in the UK, Europe, Asia and the Americas. Citizens in these communities have united to end their dependency on fossil fuels. The Transition Network provides a 12-step guide to a low carbon economy (Transition Network.org/) http://www.transitionnetwork.org/).

5. Eco-Cities. The largest energy savings will come from replacing single-driver cars by redesigning cities into small clusters designed for close access so that people will be able to answer all their needs within the radius of a five-minute walk. The California Infill Builders Association, (CIBA) anticipates that intermingling housing, small commerce, and workplaces would cut driving miles in California by half and save $4.3 billion in infrastructure costs. Such mixed use neighborhoods would reduce commuting by 3.7 trillion miles—the equivalent of removing every car from the state's roads for 12 years, and save the average household $6,500 in reduced car and utility expenses (see http://infill-builders.org/).

With specific reference to nuclear plants, the less wattage consumed, the less power is needed; by reducing energy demand, nuclear power becomes irrelevant, and the argument for shutting down all 104 nuclear plants in the United States, and refusing to build any new ones, regardless of how much profit politicians and corporations stand to make, becomes that much stronger (Gar Smith. *Nuclear Roulette,* 62).

On Christmas day, Day 289, Stefan Schurig, Climate Energy Director of the World Future Council, brings some joy to the world: "Fukushima is one of the last nails in the coffin for nuclear energy...[in terms of raising] doubts about the safety of nuclear energy." Schurig is confident that a non-nuclear future is ahead of us: "the catastrophe at Fukushima exposed nuclear power's cost and potential shortfalls. It's only a matter of time before all countries phase out nuclear power" (Calthorpe Associates, "Vision California: Charting Our Future").

129.

Day 295. Envoie

Bonnie Urfer is still in jail, serving out her eight-month sentence. Writing in the Summer edition of *Nukewatch* quarterly she has this to say:

> I have worked at *Nukewatch* for 25 years and cannot stop working against nuclear power and weapons. I have petitioned, written letters, protested, and gone to jail many times and will again soon for opposing nuclear weapons production at Y-12 complex in Tennessee. Nonviolent resistance at the risk of arrest and prison is a sane response to a callous government and greedy industry, yet I regret and apologize for not being more effective.
>
> I have long thought that politics of the heart have never been able to keep pace with politics of the mind. Imagine what the world would look like if for the past 50 years all of the trillions of dollars spent on weapons of mass destruction and nuclear power would have been poured into renewable energy sources instead. Imagine how life affirming policies could have enhanced world-wide community and relief from the suffering that is continuously caused by war and war preparations; if instead of politics of the mind focused on domination and profit, we had collectively embraced politics of the heart, focusing on what is best for us all, now and for future generations.
>
> *Believe me when I say that those of us who have been doing antinuclear work for decades desperately need help. Every person is needed to commit to ending the nuclear age—today, in our lifetime—to lessen the impact on every future generation. There is no other solution.*

Perhaps it is not given on this Earth to learn all the marching steps, and those we have learned, we have not yet learned fast enough. Not even close. Not in the time left to us. At best we can learn to place one foot before another while we fathom the great divide: "Why should one man feast while another man starve?" Tom Paine asked his question as a great empire writhed in its birth pangs. His sons and daughters, the 99%, are asking it now.

What is right? What right have we to do what is right? What right remains to us, knowing the little that we know? Setting one foot forward before the other, we will put our solar panels on our roofs, we will drive less, or not at all, we will walk a little more. We will learn to share. We will own our workplaces, we will form our state and city banks, we will chain ourselves to reactor gates with our signs and our puppets, and our drums and fifes. We will risk arrest. We will circulate anti-nuclear ballot initiatives. We will appear in the chambers of our local municipalities, and march in the streets, we will speak what words our insect tongues can generate that day. We will live the slow resistances of which we are capable. We are what we are.

Sometimes, in times of deep despair, I walk in the forest. I set myself a task. Every tree, every leaf, every flutter of birdwing in the branches, every stone, every pebble in the path, all these I see knowing that if I fail to see, that single thing will disappear, will vanish from the Earth, to be sucked up by nothingness, never to return.

Only an exercise, a letting out of the mind's horse from its stable for a slow run in the What If?

Will you join me?

130.

What the Light Was Like

The ancient Mexicas, my ancestors, believed that if you are chosen to be loved by a particular tree, that love is the greatest love there is; it is a love that is boundless, timeless and filled with the tenacious spirit that roots it to the earth's soil. It is the spirit guide that accompanies you wherever you walk the earth.

I sit on the stone threshold of a farmhouse, slightly hollowed now by the come and go of peasant steps. I feel the soft grass beneath my feet. On days like this you can see as far as Salzburg, where the Dachstein raises its shadow against the brightness. All around to the far horizon, the hills spool out their waves of grass. A country road runs below the bottoms, out of sight except for the rare car passing. There's the still of the early afternoon, as if the air itself listened to its own breathing, and the slant of sunlight riffling each blade of grass rouses the dumb earth to sing in shimmering harmonies.

I mark the hours of daylight drifting through the trees, watching the light ease itself around their trunks, envelop first one, then another, until it fires the branches—forty-three trees in this orchard—fanned in rows on this mounded hill as far as eye can reach. There is no sense of time, no time any longer, just the shifting play of light, veiling, now unveiling, folding, now unfolding about each branch, each trunk, the daily play of light, each day different as the sun's course shifts through summer into autumn (not so imperceptibly as to be imperceptible) as each hour passes, as each trunk, and branch, and leaf and fruit, becomes Tree, and Branch and Leaf, confirmed in light.

Light gives you this name, name of tree, name of branch, of leaf, of ripening fruit—this fruit and no other—clusters red to bursting, this nectar weeping sweetness through burst skins.

If I close my eyes now, I can hear the hum of bees. Twenty-five years later, I can see the orchard, the soft light easing over the bark, enveloping each trunk in reflected light, until each passes into the ranks of its own perfection. Light in the grass, in the leaves, the air listening to itself breathe. The still of light-mirroring-light, the way time stops for the opening eye, the listening ear.

This was Earth, the way we lived it, felt its presence, felt the stir within our bones, our bones with it and of it. If I could tell you—if you could hear, in the age that comes after—what it was like to live in this Fourth Sun. In its perfection there was everything.

131.

Epilogue: Quid Pro Quo

• On Day 462 PM, Noda, yielding to pressure from the nuclear industry, gave Kansai Electricity the go-ahead to restart Oohi reactors 3 and 4, triggering a Japan-wide outcry, with 200,000 protesters overcoming their cultural restraints to clog the streets of Tokyo every Friday night in an on-going demonstration that shows no sign of letting up.

• On Day 558, the U.S. government announced that it would station a new defense radar system in Japan "to help protect the country." The U.S. does not admit that the gift is related in any way to encircling China, or to pressuring Japan to totally reverse PM Noda's recent announcement that Japan would achieve a complete nuclear phase-out by 2030 (SimplyInfo. ""Japan's Nuclear Program: Rearranging Deck Chairs on the Titanic.")

• Russia's nuclear industry is pitching low-cost reactor deals to foreign buyers citing Chernobyl as a selling point. (Eve Conant. "Russia Uses Lessons of Chernobyl as a Selling Point for its Reactor Technology.")

• As of Day 571 following the nuclear meltdowns at Fukushima Daiichi, although TEPCO has capped Unit 4 with a 60-ton roof, earthquake damage to its structural integrity remains of critical concern to the entire planet.

WARNING

In the jungle of my mind, one stele remains. For those who can decipher, it reads:

When the north star will appear to have shifted 40 degrees in the constellation of the bear, you will come here. You will not see the same stories we saw in the night sky. You will not learn from our mistakes. There is no way to warn you. Our writing will have faded, our stone monuments will have turned to dust, our inscriptions will yawn emptiness. We cannot imagine your journey, or your way of being. You will be living things, possessed as we were of ways of seeing, ways of hearing. They will not be our ways. They will be other ways.

Treat her well, this Earth. Love her in her infinite beauty. Let her feed you what fruits you need. If you let her, she will give to you. Her soil will nurture you and lay you to rest. She lives in you, in your marrow and sinew. Remember. You live in her. Love her as your mother. Be kind. Do not take from her. Do not fail to recognize her gifts. Remember, as we did not, she is your mother. Do not abandon her as we did.

Appendix 1: Nuclear Information Resources

Beyond Nuclear
6930 Carroll Avenue, #400
Takoma Park, Maryland 20912
(301) 271-2209
Web: beyondnuclear.org • E-mail: info@beyondnuclear.org

Centre d'Enseignement et de Recherche en Environnement Atmospherique
6-8 avenue Blaise Pascal, Cité Descartes Champs-sur-Marne
77455 Marne la Vallée Cedex 2
Téléphone : +33 1 64 15 21 57 - Fax : +33 1 64 15 21 70
http://cerea.enpc.fr/fr/ • E-mail: wmcerea@cerea.enpc.fr

Ctizens Awareness Network (CAN)
P.O. Box 83
Shelburne Fall, MA 01370
(413) 339-5781
www.nukebusters.org

Ecological Options Network (EON)
www.eon3.net or EON3emfBlog.net
EON's YouTube Channel
PlanetarianPerspectives.net

Electronic Privacy Information Center
1718 Connecticut Avenue, NW #200
Washington, D.C. 20009
(202) 483-1140
Web: epic.org

Federation of American Scientists
1725 DeSales Street, NW, 6th floor
Washington, D.C. 20036
(202) 546-3300
Web: fas.org • E-mail: fas@fas.org

The Global Network Against Weapons and Nuclear Power in Space
Bruce K. Gagnon, Secretary
P.O. Box 652
Brunswick, ME 04011
(207) 443-9502
www.space4peace.org • E-mail: globalnet@mindspring.com

Hanford Watch
Portland, Oregon
http://www.hanfordwatch.org/ • E-mail: paigeknight@comcast.net

Institute for Energy & Environmental Research
6935 Laurel Avenue, No. 201
Takoma Park, Maryland 20912
(301) 270-5500
Web: ieer.org • E-mail: info@ieer.org

Institute for Policy Studies
1112 16th Street NW #600
Washington, D.C. 20036
(202) 234-9382
Web: ipc.org • E-mail: info@ipc.org

International Campaign to Ban Uranium Weapons
ICBUW.org
Bridge 5 Mill
22a Beswick Street
Ancoats
Manchester
M4 7HR
UK
Tel/Fax: +44 (0) 161 2738293
http://web.bandepleteduranium.org/tools/form.php?id=80&id_topic=1

Los Alamos Study Group
2901 Summit Place, NE
Albuquerque, New Mexico 87106
(505) 265-1200
Web: lasg.org
E-mail: gmello@lasg.org, or twm@lasg.org

Low Level Radiation Campaign
Powys LD1 5LW, UK
Tel. 44-1597-824-771
Web: llrc.org • E-mail: sitemanager@llrc.org

Midwest Renewable Energy Association
7558 Deer Road
Custer, Wisconsin 54423
(715) 592-6595
Web: midwestrenew.org • E-mail: info@the-mrea.org

NCWARN
North Carolina Waste Awareness & Reduction Network
P.O. Box 61051
Durham, NC 27715-1051
(919) 416-5077
www.ncwarn.org • E-mail: ncwarn@ncwarn.org

New England Coalition on Nuclear Pollution (NECNP)
P.O. Box 545
Brattleboro, VT 05302
(802) 257-0336
www.newenglandcoalition.org • E-mail: necnp@necnp.org

NJPIRG Citizen Lobby
143 East State Street, Suite 6
Trenton, NJ 08608
(609) 394-8155
www.njpirg.org

Nuclear Energy Information Service (NEIS)
3411 W. Diversity Avenue, #16
Chicago, IL 60647
(773) 342-7650
www.neis.org • E-mail: neis@neis.org

Nukefree.org
Harvey Wasserman, Sr. Advisor; E-mail: windhw@mac.com
Mary Skerrett, Director; E-mail: mary@nukefree.org
www.nukefree.org

Nuclear Information & Resource Service
6930 Carroll Avenue #340
Tacoma Park, Maryland 20912
(301) 270-6477
Web: nirs.org • E-mail: nirsnet@nirs.org

Nuclear Age Peace Foundation, NY
446 East 86 St.
New York, NY 10028
(212) 744-2005
(646) 238-9000(cell)
Web: www.wagingpeace.org or www.abolition2000.org

NuclearResister
P.O. Box 43383
Tucson, Arizona 85733
(520) 323-8697
E-mail: nukeresister@igc.org

Nukewatch
The Progressive Foundation
740 A Round Lake Road
Luck, Wisconsin 54853
E-mail: nukewatch1@lakeland.ws

Physicians for Social Responsibility
1875 Connecticut Avenue, NW
Washington, D.C. 20009
(202) 667-4260
Web: psr.org
Wisconsin Chapter
P.O. Box 1712
Madison, Wisconsin 53701
(608) 232-9945
E-mail: info@psrwisconsin.org

Public Citizen
1600 20th Street, NW
Washington, D.C. 20009
(202) 588-1000
Web: citizen.org
E-mail: member@citizen.org

Radiation and Public Health Project
Joseph Mangano
Executive Director
P.O. Box 1260
Ocean City, NJ 08226
609 399-4343
email: odiejoe@aol.com

Riverkeeper
828 South Broadway
Tarrytown, NY 10591
(914) 478-4501
e-mail: info@rierkeeper.org
www.riverkeeper.org

Rocky Mountain Institute
1820 Folsom Street
Boulder, Colorado 80302
(303) 245-1003
Web: rmi.org
E-mail: engage@rmi.org

Southern Alliance for Clean Energy (SACE)
P.O. Box 1842
Knoxville, TN 37901
(865) 637-6055
e-mail: info@cleanenergy.org
www.cleanenergy.org

Three Mile Island Alert
4100 Hillsdale Road
Harrisburg, PA 17112
(717) 541-1101
e-mail: tmia@tmia.com
www.tmia.com

Union of Concerned Scientists
2 Brattle Square
Camrbridge, Massachusetts 02138
(617) 547-5552
Web: ucsusa.org

Western States Legal Foundation
655 – 13th Street, Suite 201
Oakland, CA 94612
(510) 839-5877
Web: wslfweb.org • E-mail: wslf@earthlink.net

Appendix 2: Blogs

Abalone Alliance Safe Energy — http://www.bapd.org/gab-se-1.html

Akio Matsumura.com — http://akiomatsumura.com/

The Carbon Capture Report — http://www.carboncapturereport.org/

Capitoilette — http://capitoilette.com/

DC BUREAU — http://www.dcbureau.org/

Depleted Cranium — http://depletedcranium.com/best-sources-for-information-on-the-fukushima-nuclear-reactors/

ENE news — http://enenews.com/

The Energy Net — http://www.energy-net.org

Firedoglake — http://news.firedoglake.com/

Fukushima Green Action — http://fukushima.greenaction-japan.org/

Greenpeace — http://www.greenpeace.org/international/en/news/Blogs/makingwaves/fukushima-latest-update-and-where-to-get-more/blog/33757/

Global Research, Center for Research on Globalization — http://www.globalresearch.ca/index.php?context=home

National Resource Defense Council — http://www.nrdc.org/globalwarming/

NEI Nuclear Notes — http://www.neinuclearnotes.blogspot.com/

Nuclear Resource and Information Service — http://www.nirs.org/

NukNews — http://groups.yahoo.com/group/NucNews/

The Carbon Capture Report — http://www.carboncapturereport.org/

The Fukushima Project — http://www.simplyinfo.org/?p=3388

Appendix 3: Activist Organizations

Abolition 2000 Global Network to Eliminate Nuclear Weapons
Web: abolition2000.org

Alliantaction.org
Eden Prairie, Minnesota
http://www.alliantaction.org/home.html
e-mail: alliantaction@circlevision.org

Blue Ridge Environmental Defense League
A six-state SE regional group
PO Box 88
Glendale Springs, NC 28629
Phone: (336) 982-2691
Fax: (336) 982-2954
email:bredl@skybest.com
http://www.bredl.org/contactus.htm
http://www.bredl.org/nuclear/index.htm
http://www.bredl.org/nuclear/Bellefonte.htm
http://www.bredl.org/nuclear/Vogtle.htm
http://www.bredl.org/nuclear/NorthAnna.htm
http://www.bredl.org/nuclear/WSLee.htm

Georgia WAND
Georgia Womens' Action For New Directions
250 Georgia Ave. SE, Ste. 202
Atlanta, GA 30312
404-524-5999
404-524-7593 FAX
email: georgiawand@wand.org

The Guacamole Fund
P.O. Box 699
Hermosa Beach, CA 90254
(310) 374-4837
e-mail: guacamaole@bigplanet.com
www.guacfund.org

Hanford Watch
Portland, Oregon
http://www.hanfordwatch.org/

International Campaign to Ban Uranium Weapons
Web: bandepleteduranium.org

International Physicians for the Prevention of Nuclear War
Web: ippnw.org

NUCLEAR FREE CALIFORNIA
A major coalition of California Groups
http://www.nuclearfreecal.org/nfcnet/

The Nuclear World Project
Web: thenuclearworld.org

NUKEFREE.ORG
Harvey Wasserman
Senior Advisor & Wedbsite Editor:
e-mail: windhw@mac.com
Mary Skerrett
Program Director and Outeach Coordinator
e-mail: mary@nukefree.org
www.nukefree.org

Peace Boat
Web: peaceboat.org

Peace Works
Kansas City, Missouri
http://www.peaceworkskc.org/kcplant
e-mail: http://www.peaceworkskc.org/contact.html

Three Mile Island Alert
4100 Hillsdale Road
Harrisburg, PA 17112
(717) 541-1101
e-mail: tmia@tmia.com
www.tmia.com

SafePower Vermont
http://vermont.sierraclub.org/contact.html

SAGE Alliance
A Coalition of Vermont Groups
http://sagealliance.net/who_we_are/regional_groups

Women's Energy Matters,
a California organization working for a sustainable energy system without nuclear power.
http://www.womensenergymatters.org

WIKIPEDIA
List of anti-nuclear groups in USA
http://en.wikipedia.org/wiki/Anti-nuclear_groups_in_the_United_StatesAPPENDIX I

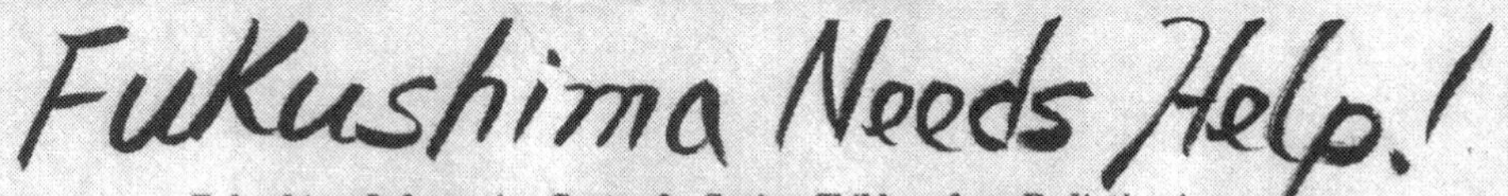

Fukushima Information Center for Saving Children from Radiation / Citizens' Radioactivity Measuring Station

After March 12th, people who reside in Japan have no choice but to live with radiation. As there is no international consensus on a safe threshold level for radiation exposure, it must be avoided and be reduced by individual efforts. The radiation has not been contained within a certain radius or within the borders of local governments. The Citizens' Radioactivity Measuring Station was founded in response to this situation. It measures the radiation for citizens' self-protection and provides the tools for acquiring more knowledge of radiological defense to facilitate individuals making their own decisions.

From Fukushima:
Dear friends,
We are in Fukushima city just 60km (37 miles) away from the Fukushima nuclear plant.
Now, our children's physical condition is in danger.
We see symptoms such as swollen thyroid, nosebleed, diarrhea, cough, asthma, etc.

We, Mothers of those children, believe it is a very serious situation and worry about the effects of atomic radiation on our children's health,
Our government has been announcing that there is no effect on our health from radiation at all.
We cannot believe our government's safety standard any more.
We cannot wait our children to have a cancer from the high level of exposure.
We have decided to start up an organization in order to protect our children.
We are planning to research on health issues by ourselves with the cooperative scholars and experts with our own independent network.
Please help us protecting our children from the effect of radiation for our and your future!!
Sincerely,
Aya Marumori,
Representative of FUKUSHIMA Information Center for Saving Children from Radiation

SEND DONATION (International Postal Money Order) TO:

SHIMIN HOSHANO SOKUTEISHO (Citizens' Radioactivity Measuring Station)
〒960-8036 8-8 Shinmachi, Fukushima-shi, Meiho bld.3rd floor, FUKUSHIMA JAPAN
http://en.crms-jpn.com/ (English) http://www.crms-jpn.com (Japanese)

http://blog.goo.ne.jp/kodomofukushima

info@crms-jpn.com

BIBLIOGRAPHY

BOOKS

Abley, Mark. *The Prodigal Tongue: Dispatches from the Future of English.* NY: Houghton Mifflin, 2008.

Abley, Mark. *Spoken Here: Travels among Threatened Languages.* NY: Houghton Mifflin, 2003.

Alexievich, Svetlana. *Voices from Chernobyl.* Trans. Keith Gessen. Normal, IL: Dalkey Archive Press, 2005. Page citations are from *Voices from Chernobyl: The Oral History of a Nuclear Disaster.* NY: Picador, 2006. First published in Russian as *Tchernobylskaia Molitva* by Editions Ostojie, 1997.

Athanasiou, Tom. *Divided Planet: The Ecology of Rich and Poor.* Boston: Little Brown & Co., 1996.

Bennet, Jeffrey, Meagan Donahue, *et al. The Cosmic Perspective.* San Francisco: Pearson/Addison-Wesley, 2007.

Beradt, Carlotte. *The Third Reich of Dreams.* NY: Quadrangle Books, 1968.

Berger, John. *Hold Everything Dear: Dispatches on Survival and Resistance.* NY: Pantheon, 2007.

Bertell, Rosalie. *No Immediate Danger.* London: The Women's Press, 1985.

Grace Bukowski, Damacio A. Lopez, and Fielding M. McGehee III. "Uranium Battlefields Home & Abroad: Depleted Uranium Use by the US Department of Defense." Reno and Carson City, NV: Rural Alliance for Military Accountability, Progressive Alliance for Community Empowerment, and Citizen Alert, March 1993.

Caldicott, Helen. *Nuclear Power is Not the Answer.* NY: New Press, 2006.

Carlin, George. *Last Words.* NY: Free Press, Simon & Schuster, 2009.

D'Gata, John. *About a Mountain.* NY: W.W. Norton & Co., 2010.

Glendinning, Chellis. *Waking Up in the Nuclear Age.* Philadelphia: New Society Publishers, 1987.

Gofman, John W. *An Irreverent Illustrated View of Nuclear Power.* San Francisco: Committee for Nuclear Responsibility, 1979.

Greene, Gayle. *The Woman Who Knew Too Much: Alice Stewart and the Secrets of Radiation.* Ann Arbor, MI: U. Michigan Press, 1999.

Hinton, Leanne. *Flutes of Fire: The Indian Languages of California*. Berkeley: Heyday Books, 1993.

Hoffman, Ace. *The Code Killers: Why DNA and Ionizing Radiation are a Dangerous Mix. An Exposé of the Nuclear Industry.* Available for free download at www.acehoffman.org, 2008.

Johnston, Barbara Rose, ed. *Half Lives & Half Truths: Confronting the Radioactive Legacies of the Cold War.* Santa Fe: School for Advanced Research Press, 2007.

LaConte, Ellen. *Life Rules,* New York, iUniverse, 2010.

League of Women Voters. *Nuclear Waste Primer: A Handbook for Citizens.* NY: Lyons & Burford, 1993.

Le Pen, Jany. *SOS: Enfants d'Irak,* Paris: Editions Objectif, 2001.

Lilly, John. *Lilly on Dolphins: Humans of the Sea.* NY: Anchor Press, 1975.

— *Man and Dolphin: Adventures of a New Scientific Frontier*. Garden City, NY: Doubleday, 1961.

— *Man and Dolphin: The Possibilities of Talking with other Species*. NY: Julian Press, 1987.

— *The Mind of the Dolphin: A Non-Human Intelligence.* Garden City, NY: Doubleday, 1967.

Macy, Joanna. *Despair and Personal Power in the Nuclear Age.* Philadelphia: New Society Publishers, 1983.

Makhijani, Arijun, and Scott Saleska: *The Nuclear Power Deception: U.S. Nuclear Mythology from Electricity "Too Cheap to Meter" to "Inherently Safe" Reactors.* NY: The Apex Press, 1999.

McBay, Aric, Lierre Keith, and Derrick Jensen. *Deep Green Resistance*. NY: Seven Stories Press, 2011.

Manzurova, Natalia. *Hard Duty: A Woman's Experience at Chernobyll.* Trans. Cathie Sullivan. 64 pages. Ozersk, Russia: self-published, n.d.

Mouchot, Augustin. *Chaleur Solaire: Applications Industrielles.* Paris: Imprimerie Ernest Mazerai, 1869.

Nader, Ralph, and Jon Abbotts. *The Menace of Atomic Energy.* NY: W. W. Norton & Co, Inc., 1977.

Nollman, Jim. *The Charged Border*. NY: Henry Holt & Co., 1999.

Palast, Greg. *Vultures' Picnic: In Pursuit of Petroleum, Pigs, Power Pirates, and High Finance Carnivores*. NY: Dutton, 2011.

Rumi, Jelaluddin. *The Essential Rumi.* Trans. Coleman Bark. Edison, NJ: Castle Books, 1997.

Schcherbak, Iurii. *Chernobyl: A Documentary Story.* Trans. Ian Press. London: Macmillian Press Ltd, 1989.

Scheer, Hermann. *Energy Autonomy.* Trans. Jeremiah M. Reimer. Sterling, VA: Earthscan, 2007.

Schell, Jonathan. *The Fate of the Earth.* NY: Alfred A. Knopf, 1983.

Smith, Gar. *Nuclear Roulette.* San Francisco: International Forum on Globalization, 2011.

Wasserman, Harvey, Norman Solomon, *et al. Killing Our Own: The Disaster of America's Experience with Atomic Radiation.* NY: Dell, 1982.

Wilkerson, Isabel. *The Warmth of Other Suns.* NY: Random House, 2010.

ARTICLES

Abalone Alliance Clearing House. "Fukushima FOIA: The Aftershock [Quake] That Blew Up Unit 1." *Hawaii News Daily,* October 23, 2011. http://hawaiinewsdaily.com/2011/10/fukushima-foia-the-aftershock-that-blew-up-unit-1/

Al-Azzawi, Souad N. "Depleted Uranium Radoactive Contamination In Iraq: An Overview." *Brussels Tribunal,* August, 2006. http://www.brussels-tribunal.org/DU-Azzawi.htm

—. "The Responsibility of the U..S. in Contaminating Iraq with Depleted Uranium." *Global Research.* November 8, 2009. http:www.globalresearch.ca/index.php?context=va&aid=15966

Al Jazeera English. "Fukushima Radiation Alarms Doctors," *Al Jazeera English*, Aug. 18, 2011. http://www.aljazeera.com/indepth/features/2011/08/201181665921711896.

Calthorpe Associates. "Vision California: Charting Our Future." http://www.visioncalifornia.org/reports.php

Chulov, Martin. "Research Links Tie Falluja Birth Defects and Cancers to US Assault." *Guardian UK,* December 30, 2010. http://www.guardian.co.uk/world/2010/dec/30/falluja-birth-defects-iraq

Clenfield, Jason. "Vindicated Siesmologist Says Japan Still Underestimates Threat to Reactors." *Bloomberg News*, Nov. 21, 2011. http://www.bloomberg.com/news/2011-11-21/nuclear-regulator-dismissed-seismologist-on-japan-quake-threat.html

Cohen-Joppa, Felice, and Jack Cohen-Joppa (eds). "Vermont Yankee." *The Nuclear Resister*, June 7, 2011. http://www.nukeresister.org/wp-content/uploads/2011/06/NR162.pdf

Conant, Eve. "Russia Uses Lesson of Chernobyl as a Selling Point for Its Reactor Technology." http://www.scientificamerican.com/article.cfm?id=russia-uses-lesson-of-chernoby-as-selling-point-for-its-reactor-technology

Dangelis, Alyssa. "Solar-Powered Robot Swarm Could Clean Oil." *Discovery News*, Aug. 27, 2010. http://news.discovery.com/tech/solar-powered-ocean-robots-clean-oil.html

Doan, Abigail. "Green Building in Zimbabwe Modeled on Termite Mounds." *Inhabitat*, Dec. 10, 2007. http://inhabitat.com/building-modelled-on-termites-eastgate-centre-in-zimbabwe/

Donn, Jeff. "AP Impact: Industry and NRC Re-Write Nuke History." *ABCNews*, June 28, 2011. http://abcnews.go.com/US/wireStory?id=13945567

Dupré, Deborah. "Obama's Dirtiest Secret Exposed: Plutonium 'Bomb Plant,' 'Green Future.'" *Examiner.com*, Sep. 14, 2001. http://www.examiner.com/human-rights-in-national/obama-s-dirtiest-deadly-secret-exposed-plutonium-bomb-plant-green-future

Editors, "German Radiation Professor Warns of Possible Nuclear Explosion at Fukushima." *Energy News*, Nov. 6, 2011. http://enenews.com/just-in-german-nuclear-professor-fukushima-was-a-kind-of-mini-atomic-bomb-warns-against-further-nuclear-explosions. [English abstract of "Deutscher Strahlen-Experte warnt: Fukushima ist Atombombe im Mini-Maßstab." *Bild*, Nov. 5, 2011. http://www.bild.de/news/ausland/fukushima/fukushima-wie-mini-atombombe-20843412.bild.html]

— "Report Confirmed That the Wall of Reactor 4 was Lost on the South Side." *Energy News*, Dec. 12, 2011. http://enenews.com/report-confirmed-wall-reactor-4-lost-south-side-photos

— "Diseased Alaska Seals Tested for Radiation." *Energy News*, Dec. 28, 2011. http://enenews.com/diseased-seals-abnormal-brain-growths-under-sized-lymph-nodes-found-russia-canada-walruses-next-6-photos-map

Fackler, Martin. "In Nuclear Crisis, Crippling Mistrust." *New York Times*, June 12, 2011. http://www.nytimes.com/2011/06/13/world/asia/13japan.html?pagewanted=all

George, Barbara. Women's Energy Matters: Opening Brief in Track 1 – Local Capacity Resources, 2011

Grossman, Karl. "Fukushima and the Nuclear Establishment." *Counterpunch*, June 16, 2011. http://www.counterpunch.org/2011/06/16/fukushima-and-the-nuclear-establishment/

Jarvis, Kyle. "A Federal Court Gives Vermont Yankee the Go-Ahead." *The Keene Sentinel*, Jan. 20, 2012. http://www.sentinelsource.com/news/local/a-federal-court-gives-vermont-yankee-the-go-ahead/article

Kakuchi, Suvendrini. "Mothers Rise Against Nuclear Power." *Common Dreams*, Dec. 22, 2011. http://www.commondreams.org/headline/2011/12/22-1

Levine, Gregg. "NRC Chair Jaczko: Event Like Fukushima Too Rare to Require Immediate Changes." *Firedoglake*, Oct. 10, 2011. http://my.firedoglake.com/gregglevine/2011/10/11/nrc-chair-jaczko-events-like-fukushima-too-rare-to-require-immediate-changes/

— "Gregory Jaczko Has a Cold," *Capitoilette*, Dec. 9, 2011. http://capitoilette.com/2011/12/09/gregory-jaczko-has-a-cold/

Mainichi Daily News. "U.S. Nuclear Chief Says Nuclear Plant Stable But Major Task Remains," *Mainichi Daily News*, Jan. 19, 2012. http://www.silobreaker.com/us-nuclear-chief-says-fukushima-plant-stable-but-major-task-remains-5_2265076045075972149 [webpage at mainichi.jp no longer available]

Matsumura, Akio. "The Fourth Reactor and the Destiny of Japan." *AkioMatsumura.com*, Sep. 29, 2011. http://akiomatsumura.com/2011/09/the-fourth-reactor-and-the-destiny-of-japan.html

McNeill , David, and Jake Adelstein. "The Explosive Truth behind Fukushima's Meltdown." *The Independent*, UK, Aug. 17, 2011. http://www.independent.co.uk/news/world/asia/the-explosive-truth-behind-fukushimas-meltdown-2338819.html

Mubayi, V, et al. "Cost-Benefit Considerations in Regulatory Analysis." Brookhaven National Laboratory. http://pbadupws.nrc.gov/docs/ML1030/ML103050362.pdf

Nuclear Information and Resource Service (NIRS). "Fact Sheet on Fukushima Nuclear Power Plant." http://www.nirs.org/reactorwatch/accidents/Fukushimafactsheet.pdf

Nuclear Resister. "Global Resistance to Nuclear Power on the Rise." June 7, 2011. http://www.nukeresister.org/wp-content/uploads/2011/06/NR162.pdf

Obe, Mitsuru. "No Error in Nuclear Crisis." *The Wall Street Journal*, Dec. 3, 2011. http://online.wsj.com/article/SB10001424052970204012004577073492586384110.html

Oda, Mayumi. "The Two-Headed Monster of Poison Fire." *The Safe Energy Handbook*, by Jan Thomas, Claire Greensfelder, and Wendy Osler, with Nora Akino. INOCHI/Plutonium Free Future, 1997.

O'Keefe, Derrick. "After Durban: We Must Pull the Emergency Brake Before the 1 Per Cent Drive Us Off the Cliff." *Common Dreams*, Dec. 18, 2011. http://www.commondreams.org/view/2011/12/18-1.

Redman, Janet. "Connecting the Climate Dots Across the Map." *AlterNet*, July 18, 2011. http://www.alternet.org/story/151671/connecting_extreme_weather_dots_across_the_map/

Sherman, Janette D., MD, and Joseph Mangano. "Is the Dramatic Increase in Baby Deaths in the U.S. a Result of Fukushima Fallout?" *Counterpunch*, June 12, 2011. http://www.counterpunch.org/2011/06/10/is-the-increase-in-baby-deaths-in-the-us-a-result-of-fukushima-fallout/

— "An Unexpected Mortality Increase In The United States Follows Arrival Of The Radioactive Plume From Fukushima: Is There A Correlation?" *International Journal of Health Services* 42.1 (2012), 47-64.

SimplyInfo. "Reporter Works at Fukushima, Exposes Real Situation at the Plant." *SimplyInfo*, Dec. 17, 2011. http://www.simplyinfo.org/?p=4349

— "Rearranging Deck Chairs on the Titanic." *SimplyInfo*, Sept. 20, 2012. http://www.simplyinfo.org/?p=7524

Tabuchi, Hiroko. "Japan's Prime Minister Declares Fukushima Plant Stable." *New York Times*, Dec. 16, 2011. http://www.nytimes.com/2011/12/17/world/asia/japans-prime-minister-declares-fukushima-plant-stable.html

Trento, John. "MOX Fuel Rods Used in Japanese Reactor." *DC Bureau*, Mar. 15, 2011. http://www.dcbureau.org/20110315782/natural-resources-news-service/mox-fuel-rods-used-in-japanese-nuclear-reactor-present-multiple-dangers.html

UN Chernobyl Forum Expert Group 'Environment.' "Environmental Consequences of the Chernobyl Accident and Their Remediation: Twenty Years of Experience." Vienna: The International Atomic Energy Agency, 2006. http://www-pub.iaea.org/mtcd/publications/pdf/pub1239_web.pdf

Viewzone. "Depleted Uranium: Anatomy Of An Atrocity." *Viewzone*. http:://www.viewzone.com/du/du22.html

WACEB. "Nuclear Expert: 'Unit 4 is Looking More and More Like the Leaning Tower of Pisa Right Now'", *WeAreChangeTV.us*, Dec. 20, 2011. http://wearechangetv.us/2011/12/nuclear-expert-unit-4-is-looking-more-and-more-like-the-leaning-tower-of-pisa-right-now/#ixzz1jsMEYx5E

Wald, Matthew L. "Approval of Reactor Design Clears Path for New Plants." *New York Times*. Dec. 22, 2011. http://www.nytimes.com/2011/12/23/business/energy-environment/nrc-clears-way-for-new-nuclear-plant-construction.html

Wasserman, Harvey. "Where Occupy and No Nukes Merge and Win." *Common Dreams,* Oct. 24, 2011. http://www.commondreams.org/view/2011/10/24-5

Westerman, Doug. "Depleted Uranium—Far Worse Than 9-11." *Global Research,* May 3, 2006. http://www.globalresearch.ca/index.php?context=va&aid=15966

Wikipedia. "Depleted Uranium." Wikipedia. http:en.wikipedia.org/depleted uranium

Wingfield , Brian, and Julie Johnsson. "New York Nuclear-Accident Evacuation Would Work, Jaczko Says." *Bloomberg Businessweek*, Dec. 10, 2011. http://www.businessweek.com/news/2011-12-10/new-york-nuclear-accident-evacuation-would-work-jaczko-says.html

Zajic, Vladimir S. "Review of Radioactivity, Military Use, and Health Effects of Depleted Uranium." http://vzajic.tripod.com/1stchapter.html#top

Zharkov. "How can NATO justify their radioactive contamination of these countries with Depleted Uranium, which many consider to be crimes against humanity?" A reader comment in *Dipnote: U.S. Department of State Official Blog*, Apr. 17, 2011, responding to Barack Obama, David Cameron and Nicolas Sarkozy, "Libya's Pathway to Peace." A Joint Op-Ed in *Dipnote*, Apr. 15, 2011. http://blogs.state.gov/index.php/entires/op_ed_obama_sarkozy_cameron/ *responding to* http://blogs.state.gov/index.php/site/entry/op_ed_obama_sarkozy_cameron

Permissions

The bottom cover image of the collapsed reactor at Fukushima (2011) appeared in *The Guardian* and is used by permission of Reuters. The top cover image of the solar panel-lined roofs of Yokohama (2009) appeared in *The Japan Times.* Used by permission.

Quotations from *Half-Lives & Half-Truths* (copyright 2007 by SAR Press), edited by Barbara Rose Johnston, are used by permission of SAR Press, Albuquerque, New Mexico. Our thanks to the individual authors who are quoted: Barbara Rose Johnston, Marie I. Boutté, Paul Barb, Laura Nader and Hugh Gusterson. Quotations from "A Call to Action" by Bonnie Urfer (*Nukewatch*, Summer 2011) are used by permission of the author. Quotations from *Vultures' Picnic: In Pursuit of Petroleum Pigs, Power Pirates, and High-Finance Carnivores* by Greg Palast, are used by permission of Penguin Group (USA), Inc. Quotations from "Gregory Jaczko Has a Cold" by Gregg Levine are used by permission of the author. Quotations from *Voices from Chernobyl: The Oral History of a Nuclear Disaster* by Svetlana Alexievich, translated by Keith Gessen are used by permission of the Dalkey Archive Press; page citations are from the Picador edition of this work. Quotations from "Where Occupy and No Nukes Merge and Win" by Harvey Wasserman are used by permission of the author. All other quotations are considered to fall within "fair use" guidelines. Our thanks to all the authors quoted in this book.

Acknowledgments

My thanks go to the many friends and colleagues who have contributed to this project with their support, suggestions, and information sharing: Jackie Cabasso and Phyllis Olin, Western States Legal Foundation; Marylea Kelly, Tri-Valley Cares; David Lochbaum, Union Concerned Scientists, Mary Beth Brangan, E.O.N, and FFAN, Hattie Nestle, Citizens Awareness Network, Japanese pen-pals, Narumi Tomida and Takako Kasuya, Megan Rice, Janet Weil, Krystyna Maliniak, Lauren Elder, Bob Gorringe, Andrey Timokhin, Maria Gilardin, Angelina Llongueras, David Mark, Ralph Johansen, Fred Gajewski, Jane Eiseley, Sydney Carson, Marlene Schoofs, Dr. Laura Morgan, Donald Goldmacher, Simon Kenrick, Webb Mealy, Rhoda Curtis, Maria Espinosa, and all the many colleagues from Nuclear Free California: Donna Warnock, Donna Gilmore, Barbara George, Roger Herreid, et al., and to the U.C. Botanical Garden, my own personal XLB, as well as Georgia WAND, Beyond Nuclear, and Nuclear Information and Research Service (NIRS), to my unflinching editor, Bryce Milligan, to Ana Quintanilla for her invaluable support, to Warren and Elizabeth McKenna, and to the planet for the beauty of her infinite perfection.

About the Author

Cecile Pineda was born in Harlem, the daughter of a scholarly Mexican father, and a Swiss-French mother. Her language of origin is French.

After some twelve years of producing and directing her own experimental theater company, Pineda began to write fiction. Her novels have been critically acclaimed, with *Face* winning the Californian Commonwealth Club's Gold Medal—a record for first fiction—the Sue Kaufman Prize, and a National Book Award Nomination. Her picaresque novel, *The Love Queen of the Amazon,* written with a NEA Fiction Fellowship, was named a Notable Book of the Year by *The New York Times.* Other novels include *Frieze*, set in 9th century India and Java; *Fishlight*, a fictional memoir of childhood, and two mononovels, *Bardo99,* in which the 20th century itself passes through a bardo state; and *Redoubt*, a meditation on gender. Her play, "Like Snow Melting in Water," set in contemporary agrarian Japan, centers on themes of displacement and ecological collapse. It will see productions in India and Thailand in 2012.

Pineda has been an anti-war activist from early life. More than ever, she has turned her attention to issues affecting the sustainability of the planet. *Devil's Tango: How I learned the Fukushima Step by Step* is Pineda's anguished dissection of the nuclear industry seen through the lens of the industrial and planetary disaster now unfolding at Fukushima Daiichi. A crazy quilt of multiple voices, pieced together day-by-day, it reflects her attempt to come to terms with Fukushima's catastrophic consequences to the planet.

Visit her web page at http://www.cecilepineda.com

Wings Press was founded in 1975 by Joanie Whitebird and Joseph F. Lomax, both deceased. Bryce Milligan has been the publisher, editor and designer since 1995. The mission of Wings Press is to publish the finest in American writing—meaning all of the Americas—without commercial considerations clouding the choice to publish or not to publish. Technically a "for profit" press, Wings receives only occasional underwriting from individuals and institutions who wish to support our vision. For this we are very grateful.

Wings Press attempts to produce multicultural books, chapbooks, Ebooks, CDs, DVDs and broadsides that, we hope, enlighten the human spirit and enliven the mind. Everyone ever associated with Wings has been or is a writer, and we know well that writing is a transformational art form capable of changing the world, primarily by allowing us to glimpse something of each other's souls. Good writing is innovative, insightful, and interesting. But most of all it is honest.

Likewise, Wings Press is committed to treating the planet itself as a partner. Thus the press uses soy and other vegetable-based inks, and as much recycled material as possible, from the paper on which the books are printed to the boxes in which they are shipped.

As Robert Dana wrote in *Against the Grain,* "Small press publishing is personal publishing. In essence, it's a matter of personal vision, personal taste and courage, and personal friendships." Welcome to our world.

Colophon

This expanded and revised edition of *Devil's Tango: How I Learned the Fukushima Step by Step,* by Cecile Pineda, has been printed on 55 pound EB "natural" paper containing a percentage of recycled fiber. Titles have been set in Cochin type, the text in Adobe Caslon type. All Wings Press books are designed and produced by Bryce Milligan.

On-line catalogue
availabl
www.wingsp

Wings Press titles
to the trade
Independent Pub
www.ipgbo
and in Eu
www.gazellebooks

Also available